Electrical Machines

Machines

An Objective and
Viva Voce Approach

Electrical Machines

An Objective and Viva Voce Approach

Ankur Mathur
BE (Electrical), MTech (Hons in Power System), LMISTE

Ex-Assistant Professor
Department of Electrical Engineering
Jodhpur Engineering College and Research Centre
Jodhpur, Rajasthan

CBS Publishers & Distributors Pvt Ltd

New Delhi • Bengaluru • Chennai • Kochi • Kolkata • Mumbai • Pune
Hyderabad • Nagpur • Patna • Vijayawada

Electrical Machines
An Objective and
Viva Voce Approach

ISBN: 978-81-239-2854-8

Copyright © Author and Publisher

First Edition: 2016

Published by Satish Kumar Jain and produced by Varun Jain for

CBS Publishers & Distributors Pvt Ltd

4819/XI Prahlad Street, 24 Ansari Road, Daryaganj, New Delhi 110 002, India.

Ph: 23289259, 23266861, 23266867   Website: www.cbspd.com

Fax: 011-23243014   e-mail: delhi@cbspd.com; cbspubs@airtelmail.in.

Corporate Office: 204 FIE, Industrial Area, Patparganj, Delhi 110 092

Ph: 4934 4934   Fax: 4934 4935   e-mail: publishing@cbspd.com; publicity@cbspd.com

Branches

- **Bengaluru:** Seema House 2975, 17th Cross, K.R. Road, Banasankari 2nd Stage, Bengaluru 560 070, Karnataka
  Ph: +91-80-26771678/79   Fax: +91-80-26771680   e-mail: bangalore@cbspd.com
- **Chennai:** 7, Subbaraya Street, Shenoy Nagar, Chennai 600 030, Tamil Nadu
  Ph: +91-44-26260666, 26208620   Fax: +91-44-42032115   e-mail: chennai@cbspd.com
- **Kochi:** Ashana House, No. 39/1904, AM Thomas Road, Valanjambalam, Eranakulam 682 018, Kochi Kerala
  Ph: +91-484-4059061-65   Fax: +91-484-4059065   e-mail: kochi@cbspd.com
- **Kolkata:** 6/B, Ground Floor, Rameswar Shaw Road, Kolkata-700 014, West Bengal
  Ph: +91-33-22891126, 22891127, 22891128   e-mail: kolkata@cbspd.com
- **Mumbai:** 83-C, Dr E Moses Road, Worli, Mumbai-400018, Maharashtra
  Ph: +91-22-24902340/41   Fax: +91-22-24902342   e-mail: mumbai@cbspd.com
- **Pune:** Bhuruk Prestige, Sr. No. 52/12/2+1+3/2 Narhe, Haveli (Near Katraj-Dehu Road Bypass), Pune 411 041, Maharashtra
  Ph: +91-20-64704058/59, 32392277   Fax: +91-20-24300160   e-mail: pune@cbspd.com

Representatives

- **Hyderabad** 0-9885175004   • **Nagpur** 0-9021734563
- **Patna** 0-9334159340   • **Vijayawada** 0-9000660880

Printed at : Swastik Packagings, 506 F.I.E. Patparganj, Delhi -92

Preface

Electrical machines is being taught as a core subject in different universities and technical institutions in India. This book has been designed to prepare students for various competitive examinations. It is a collection of a large number of questions with answers on various aspects of electrical machines. The book discusses various types of electrical equipment/machines, such as single-phase transformer, three-phase transformer, electro-mechanical energy conversion, direct current generator, direct current motor, single-phase induction motor, three-phase induction motor, synchronous generator and synchronous motor.

Thus, in brief, this book provides a good material regarding various electrical machines. I hope that this handy book will prove useful for the students of diploma and degree level courses. It will also be fruitful for the students preparing for competitive examinations. I would always welcome valuable suggestions from readers and students for making improvement in the book.

Ankur Mathur

to

my

grandparents

Acknowledgements

I am thankful to my wife, parents and brothers for supporting me at every stage while preparing of the manuscript for the book. My thanks are also due to Mr YN Arjuna, CBS Publishers & Distributors, for bringing out this edition within such a short time.

Ankur Mathur

Contents

1

Direct Current Generator

MULTIPLE CHOICE QUESTIONS

Q. 1. The DC generator parts are:

  (a) Magnetic field system
  (b) Armature
  (c) Commutator and brush gear
  (d) All of the above

Ans: (d)

Q. 2. Both stator and rotor are made up of:

  (a) Silicon steel material always
  (b) Steel–brass alloy
  (c) Ferromagnetic materials
  (d) None of the above

Ans: (c)

Q. 3. The winding, in which induction of voltage takes place, is known as:

  (a) Armature winding   (b) Field winding
  (c) Interpole winding   (d) All of the above

Ans: (a)

Q. 4. The windings, which allow the flow of current so as to produce magnetic flux is known as:

  (a) Field winding
  (b) Armature winding
  (c) Interpole winding only
  (d) Frog-leg winding

Ans: (a)

Q. 5. The purpose of outer frame (also called as yoke):
(a) It forms a part of magnetic circuit
(b) It allows the path of electric current
(c) It supports the pole core
(d) Both (a) and (c)

Ans: (d)

Q. 6. Fleming's right hand rule is applicable to:
(a) Transformer   (b) Generator
(c) Motor   (d) Actuators

Ans: (b)

Q. 7. In the rotating electrical machines:
(a) South pole exists and North pole doesn't
(b) North pole exists and South pole doesn't
(c) Alternate North and South pole exist
(d) None

Ans: (c)

Q. 8. The working principle of a direct current generator is based on:
(a) Faraday's law of electromagnetic induction
(b) Moore's law
(c) Lenz's law
(d) Kirchhoff's law

Ans: (a)

Q. 9. Direct current generator is a machine that converts:
(a) Mechanical energy to heat energy
(b) Mechanical energy to light energy
(c) Mechanical energy to electrical energy
(d) Electrical energy to mechanical energy

Ans: (c)

Q. 10. Which of these may be called as the stationary parts of a direct current generator?
(a) Yoke
(b) Bearings and bed sheet

(c) Terminal box and eye bolt

(d) All of these

Ans: **(d)**

Q. 11. Which of the followings are the various rotary parts of a DC generator?

(a) Pulley and fan   (b) Commutator

(c) Armature and shaft   (d) All of the above

Ans: **(d)**

Q. 12. Initially, the electrical energy induced in a conductor will be of .... nature.

(a) AC electrical   (b) Pulsating DC

(c) Pulsating AC   (d) None of the above

Ans: **(a)**

Q. 13. In case of a DC machine, pole core is made from:

(a) Cast iron and cast steel

(b) CRGO

(c) HRGO

(d) Rolled steel

Ans: **(a)**

Q. 14. The basic purpose of laminating the armature core in a direct current machine is to minimize:

(a) Iron loss   (b) Copper loss

(c) Only hysteresis loss   (d) Eddy current loss

Ans: **(d)**

Q. 15. The losses, which occur in the teeth of a DC machine armature are:

(a) Iron loss   (b) Copper loss

(c) Hysteresis loss   (d) None of the above

Ans: **(a)**

Q. 16. What is the role of carbon brushes in DC generator?

(a) To collect the current

(b) To provide insulation

(c) To provide safety to the system

(d) None of the above

Ans: **(a)**

Q. 17. What is the basic purpose of commutator in a direct current generator?

(a) To convert direct current supply into alternating current supply

(b) To convert alternating current into direct current

(c) To carry the current in the windings

(d) All of the above

Ans: **(b)**

Q. 18. What is the purpose of brushes in DC machine?

(a) To collect mechanical energy for load circuit

(b) To carry current to windings

(c) To collect electrical energy for load circuitry

(d) None of the above

Ans: **(c)**

Q. 19. The place where brushes are used to collect the emf is called as:

(a) Commutator (b) Rocker assembly

(c) Winding (d) Interpole winding

Ans: **(b)**

Q. 20. In the construction of DC machine, commutator segments are separated by thin layers of:

(a) Paper insulation

(b) Asbestos

(c) Mica

(d) Polyvinyl chloride

Ans: **(c)**

Q. 21. In DC machines, lap winding may be used, where we need:

(a) Less amount of voltage and more current

(b) Less amount of voltage and less current

(c) High amount of voltage and less current

(d) None of the above

Ans: **(a)**

Q. 22. **In case of small size direct current generators, yokes are made up of:**

(a) CRGO (b) HRGO

(c) Cast iron (d) None of the above

Ans: **(c)**

Q. 23. **What are the basic parts of a DC generator?**

(a) Pole coils and field coils

(b) Yoke

(c) Conductors and brushes

(d) All of the above

Ans: **(d)**

Q. 24. **The main purpose of interpoles is to minimize sparking between the brushes and commutator, when the DC machine is:**

(a) Not loaded (b) Loaded

(c) Either (a) or (b) (d) All of the above

Ans: **(b)**

Q. 25. **The emf equation of a DC generator is given by:**

(a) $E = 4.44 f\phi_m N$ (b) $E = 3.79 K_A K_d N$

(c) $E = \phi ZN/60 \times P/A$ (d) $E = 3.44 f\phi_m N$

Ans: **(c)**

Q. 26. **Direct current may be collected in a DC generator by means of:**

(a) Commutator (b) Brushes

(c) Windings employed (d) Rocker assembly

Ans: **(b)**

Q. 27. **Pole shoes are used to:**

(a) Support the field coils

(b) Increase the magnetic cross-sectional area

(c) Reduce the magnetic path reluctance

(d) All of the above

Ans: (d)

Q. 28. In rotating type electrical machines, pulsation loss occurs:

(a) In core of rotor   (b) In windings

(c) In pole shoes   (d) In the body or outer frame

Ans: (c)

Q. 29. Electric force is developed in generator by:

(a) Conversion of sound energy

(b) Conversion of heat energy

(c) Conversion of light

(d) Electromagnetic induction

Ans: (d)

Q. 30. Whenever the armature winding is allowed to rotate in a field, an emf is induced. The field is known as:

(a) Magnetic field   (b) Alternating magnetic flux

(c) Electrostatic field   (d) None of the above

Ans: (a)

Q. 31. Whenever a conductor is allowed to rotate in a magnetic field, then at which particular position, the peak voltage occur?

(a) Along $40°$ angle to the magnetic field axis

(b) In the same direction to the magnetic field axis

(c) At any of the positions

(d) Along the axis of magnetic field

Ans: (d)

Q. 32. A short circuited rectangular coil falls under gravity with the coil remaining in a vertical plane and cutting magnetic lines of force. It has acceleration, which is .... in nature?

(a) Increasing   (b) Decreasing

(c) Constant   (d) Zero

Ans: (d)

Q. 33. In a rotating electrical machine, rotational losses include:

(a) Rotor core and friction losses

(b) Friction and windage losses only

(c) Stator core, friction and windage losses

(d) Rotor losses only

Ans: (c)

Q. 34. A direct current generator may be assumed as a:

(a) Pump   (b) Rectifier

(c) Rotating amplifier   (d) All of the above

Ans: (c)

Q. 35. The basic purpose of laminating the pole shoe in a DC machine is to:

(a) Reduce iron losses

(b) Reduce copper losses

(c) Reduce eddy current losses

(d) All of the above

Ans: (c)

Q. 36. In case of a DC machine, field coils are wound:

(a) In a separate manner

(b) In the armature slots

(c) Around the poles

(d) On the yoke

Ans: (c)

Q. 37. The armature core consists of grooves or slots on its:

(a) Inner surface   (b) Outer surface

(c) Both are possible   (d) None of the above

Ans: (b)

Q. 38. The DC machine armature is made up of:

(a) Insulating material   (b) Conducting material

(c) Stainless steel   (d) Silicon steel

Ans: (d)

Q. 39. Which of the following parts are useful in emf generation in a DC generator?

(a) The coil sides, placed inside the slots of armature

(b) Commutator

(c) Brush rocker assembly

(d) All of the above

Ans: **(a)**

Q. 40. Which component of direct current generator plays a vital role in providing direct current?

(a) Commutator   (b) Windings

(c) Rocker assembly   (d) Stator

Ans: **(a)**

Q. 41. Yoke carries the flux, which is:

(a) 2/3rd of the flux per pole

(b) 1/4th of the flux per pole

(c) Same as that of flux per pole

(d) Only 1/2 of the flux per pole

Ans: **(d)**

Q. 42. Which of the windings are preferred for high voltage generation?

(a) Lap winding   (b) Frog-leg winding

(c) Drum winding   (d) Wave winding

Ans: **(d)**

Q. 43. In the direct current machines, the armature windings are kept on the rotor. It is due to the requirement for:

(a) Commutation   (b) Torque movement

(c) Voltage generation   (d) Energy conversion

Ans: **(a)**

Q. 44. Welding generator will have:

(a) Lap winding   (b) Wave winding

(c) Either (a) or (b)   (d) None of the above

Ans: **(a)**

Q. 45. Rotating electrical machines, as per their power outputs may be classified as:

(a) Small machines—having output up to nearly 0.6 kW

(b) Medium machines—having output from 0.6 kW to 250 kW

(c) Large machines—having output from 250 kW to 5000 kW

(d) All of the above

Ans: **(d)**

Q. 46. The induced current in a coil is sent towards the load circuitry by connecting the terminals of coil to two continuous and insulated rings, known as:

(a) Slip rings   (b) Wire rings

(c) Collector rings   (d) Both (a) and (c)

Ans: **(d)**

Q. 47. What is the main function of field system employed in DC machine?

(a) To create a uniform magnetic field

(b) To provide current to load circuit

(c) To produce sufficient voltage across load side

(d) All of the above

Ans: **(a)**

Q. 48. Electromagnets are preferred in comparison with permanent magnets due to:

(a) Its greater magnetic effects

(b) Its field strength regulation

(c) Its low magnetic effect

(d) Both (a) and (b)

Ans: **(d)**

Q. 49. The permeability of cast steel is nearly:

(a) Twice of cast iron   (b) Thrice of cast iron

(c) Half of cast iron   (d) Three-fourth of cast iron

Ans: **(a)**

Q. 50. The purpose of dipping the windings in an insulating varnish is to obtain:

(a) Mechanical strength

(b) Good insulating properties

(c) Stiffness

(d) All of the above

Ans: (d)

Q. 51. The air gap length in DC machine is kept nearly (For a 1 kW machine):

(a) 1.0 mm to 6 mm   (b) 1.3 mm to 5 mm

(c) 2.0 mm to 5.0 mm   (d) 3.7 mm to 4.5 mm

Ans: (a)

Q. 52. The purpose of using high grade steel in armature is:

(a) To keep hysteresis loss low

(b) To reduce eddy currents

(c) To provide good output

(d) Both (a) and (b)

Ans: (d)

Q. 53. In DC machines, the laminations must be in such a direction that they are:

(a) Perpendicular to the paths of eddy currents and parallel to flux

(b) Perpendicular to flux and parallel to the paths of eddy currents

(c) Neither (a) nor (b)

(d) Both (a) and (b)

Ans: (a)

Q. 54. Brushes are staggered in DC machine to:

(a) Prevent ridge formation on the commutator surface

(b) Provide less sparking

(c) Provide low reluctance magnetic path

(d) Provide high permeability path

Ans: (a)

Q. 55. For larger machines, which types of bearings are used?

(a) Ball bearings   (b) Roller and ball bearings

(c) Thrust bearings   (d) Pedestal bearings

Ans: **(b)**

Q. 56. Which of the DC machine parts are keyed to the shaft?

(a) Commutator   (b) Cooling fan

(c) Armature core   (d) All of the above

Ans: **(d)**

Q. 57. A 4-pole DC generator is running at 1500 rpm. What will be the frequency of current in the armature?

(a) 75 Hz   (b) 50 Hz

(c) 120 Hz   (d) 225 Hz

Ans: **(b)**

Q. 58. In case of a DC machine, commutator acts as a:

(a) Half wave rectifier

(b) Centre tapped rectifier

(c) Full wave rectifier

(d) Controlled half wave rectifier

Ans: **(c)**

Q. 59. The copper segments are generally connected to the armature conductors by means of:

(a) Insulated wire   (b) Copper lugs

(c) Brushes   (d) Rocker assembly

Ans: **(b)**

Q. 60. Copper brushes are generally preferred in those systems, where:

(a) Low voltage and higher values of currents take place

(b) High voltage and low values of currents take place

(c) High voltage and high values of currents take place

(d) Low voltage and low values of currents take place

Ans: **(a)**

Q. 61. In case of wire, Aluminium is not preferred in DC machine because:

(a) It is of high value of resistivity

(b) It is of lower value of resistivity

(c) It is costlier

(d) Needs large winding space

Ans: **(d)**

Q. 62. What must be the angle between stator and rotor field in case of a DC machine?

(a) 45°   (b) 180°

(c) 90°   (d) 30°

Ans: **(c)**

Q. 63. Another name for wave winding is:

(a) Series winding   (b) Shunt winding

(c) Parallel winding   (d) All of the above

Ans: **(a)**

Q. 64. The plane, along which no emf is induced in the armature conductors, is known as:

(a) GNP   (b) MNP

(c) CNP   (d) MGP

Ans: **(b)**

Q. 65. Which one of the followings is not the bad effect of armature reaction?

(a) It changes the GNP position

(b) It reduces the induced voltage

(c) It produces sparking on the commutator

(d) It distorts and weakens the main flux

Ans: **(a)**

Q. 66. The armature reaction may be minimized by:

(a) Using compensating winding and interpoles

(b) Using the process of giving forward lead to brushes

(c) Both (a) and (b)

(d) None of the above

Ans: **(c)**

Q. 67. The value of demagnetizing amp-turns/pole may be given as:

(a) $\dfrac{Z^2\theta}{360} \times I_d$

(b) $ZI_c[(1/2p) - (\theta/360)]$

(c) $\dfrac{Z\theta}{360} \times I_c$

(d) $\dfrac{ZI_c}{2p}$

Ans: **(c)**

Q. 68. In DC generators the MNP shifts in which direction?

(a) In the forward direction

(b) In the backward direction

(c) In any direction

(d) Remains stationary

Ans: **(a)**

Q. 69. Good commutation means:

(a) No sparking at the brushes

(b) Commutator surface remaining unaffected during continuous operation

(c) Sparking at brushes is more

(d) Both (a) and (b)

Ans: **(d)**

Q. 70. Bad commutation due to mechanical consideration includes:

(a) An increment in the current density at the trailing edge of the brush

(b) Uneven commutator surface

(c) Non-uniform brush pressure

(d) Both (b) and (c)

Ans: **(d)**

Q. 71. In DC machine, current fails to reach its full value in opposite direction?

(a) Due to applied voltage

(b) Due to reactance voltage

(c) Due to back emf

(d) Due to low value of current density in windings

Ans: **(b)**

Q. 72. Commutation may be improved by using:

(a) Interpoles or compensating winding

(b) Brushes having high value of resistance

(c) A slight lead in the forward direction

(d) All of the above

Ans: **(d)**

Q. 73. Which type of commutation is produced by armature reaction and the inductance of commutating coil?

(a) Sinusoidal commutation

(b) Linear commutation

(c) Under commutation

(d) Over commutation

Ans: **(c)**

Q. 74. In DC machine, the reactance voltage (RV) has which one of the following relation:

(a) RV is proportional to I_a and also to T_c

(b) RV is proportional to T_c and $1/I_a$

(c) RV is proportional to I_a and $1/T_c$

(d) None of the above

Ans: **(c)**

Q. 75. The interpoles are provided:

(a) In between the main poles

(b) On the bottom of yoke

(c) At the coil ends

(d) On the shaft

Ans: **(a)**

Q. 76. The basic purpose of interpoles in a DC machine is to:
(a) Improve the commutation process
(b) Reduce hysteresis loss
(c) Increase eddy current
(d) Reduce sparking

Ans: (a)

Q. 77. The compensating windings are kept:
(a) On the pole core
(b) In series with conductors
(c) On the yoke in the pole faces
(d) On the pole shoes

Ans: (c)

Q. 78. The purpose of compensating winding is:
(a) To cancel the effect of voltage induced in coil due to high fluctuations of load
(b) To reduce copper losses in windings
(c) To increase the power output
(d) To reduce the friction losses

Ans: (a)

Q. 79. Permanent magnet generator is one, in which:
(a) Electromagnets are used to generate the flux
(b) A permanent magnet is used to establish the flux in magnetic circuit
(c) No compensating winding may be employed
(d) All of the above

Ans: (b)

Q. 80. Separately excited generators are those, in which:
(a) Armature is supplied with a DC source
(b) Field coil is connected in series with armature
(c) Field winding is excited from an independent external DC source
(d) None of the above

Ans: (c)

Q. 81. Which one of the following relations is not correct regarding separately excited DC generators?

(a) $P_g = E_g I$

(b) $P_L = VI$

(c) $I_a = I_L - I_f$

(d) $I_a = I_L = I$

Ans: (c)

Q. 82. In a self-excited generator:

(a) Field winding is always connected in series with load

(b) Field winding is always connected in shunt with load

(c) Field winding is excited by the current supplied by the generator itself

(d) None of the above

Ans: (c)

Q. 83. Self-excited generators may be categorized as:

(a) Series wound generators

(b) Shunt wound generators

(c) Compound wound generators

(d) All of the above

Ans: (d)

Q. 84. Which of the following relations is/are incorrect in case of DC series generators?

(a) $P_L = VI$

(b) $V = E_g - I(R_a + R_{se})$

(c) $I_a = I_L = I_{se}$

(d) $I_a = I_f = I_L - I_{se}$

Ans: (d)

Q. 85. In case of DC machine, shunt field winding has:

(a) Many turns of thick wire

(b) Large number of turns of thin wire

(c) Less turns of either thick or thin wires

(d) None of the above

Ans: (b)

Q. 86. Compound generators are of two types, which may be:

(a) Series compound and shunt compound

(b) Cumulative compound and shunt compound

(c) Differential compound and series compound

(d) Cumulative compound wound and differential compound wound generator

Ans: **(d)**

Q. 87. In a separately excited DC generator, the field winding is connected to a separate source of DC power. It may be:

(a) Another DC generator

(b) A battery

(c) A diode or controlled rectifier

(d) Any one of the above

Ans: **(d)**

Q. 88. Which type of DC generator is most suitable as booster?

(a) Shunt generator

(b) Series generator

(c) Cumulative compound generator

(d) Differential compound generator

Ans: **(b)**

Q. 89. Which generator will be proved suitable for welding purpose?

(a) Differential compound wound generator

(b) Cumulative compound wound generator

(c) Series generator

(d) None of the above

Ans: **(a)**

Q. 90. In which DC generator, residual magnetism is not needed for building up voltage?

(a) Separately excited and series generators

(b) Booster

(c) Shunt generator

(d) Separately excited DC generator

Ans: **(d)**

Q. 91. The cumulative compound wound generator may be:

(a) Over-compounded   (b) Flat compounded

(c) Under-compounded   (d) All of the above

Ans: (d)

Q. 92. In case of DC shunt generators, the value of output voltage decreases for a moment on loading condition. It happens due to:

(a) Weakening of field   (b) Resistance of armature

(c) Overheating   (d) Inductance of coil

Ans: (b)

Q. 93. In a self-excited DC generator, one of the important condition is that some residual magnetism in the core of generator must be available:

(a) True   (b) Partly true

(c) False   (d) Partly false

Ans: (a)

Q. 94. In a DC generator, the emf generated is directly proportional to the:

(a) Pole flux   (b) Armature current

(c) Supply voltage   (d) Field current

Ans: (a)

Q. 95. In a short shunt DC generator, series field is excited by:

(a) External current

(b) Armature current

(c) Shunt and series field current

(d) Load current

Ans: (d)

Q. 96. Which of the generator gives rising _V-I_ characteristics?

(a) Compound generator   (b) Shunt generator

(c) Series generator   (d) All of the above

Ans: (c)

Q. 97. What is the basic meaning of "flashing of field" in DC generator?

(a) Development of residual magnetism by a dc source

(b) Increasing flux density by providing extra ampere turns in the field

(c) Neutralization of magnetic field

(d) None of the above

Ans: (a)

Q. 98. What will be the voltage at the terminals of DC series generator at rated rpm and no load condition?

(a) Very small amount of voltage

(b) Zero

(c) More than rated voltage

(d) Half the rated voltage

Ans: (a)

Q. 99. A generator may lose residual magnetism due to:

(a) Under excitation   (b) Vibrations

(c) Heavy excitation   (d) Heating

Ans: (d)

Q. 100. DC shunt generators are most suited for stable parallel operation due to which of the following voltage characteristics?

(a) Identical   (b) Drooping

(c) Linear   (d) Rising

Ans: (b)

Q. 101. When no load voltage is kept constant, then in which one of the following DC generator, the short circuit current will be minimum?

(a) Cumulative compound wound DC generator

(b) Differential compound wound DC generator

(c) Shunt generator

(d) Booster

Ans: (b)

Q. 102. Which one of the following generators is normally used in alternators?

(a) Shunt

(b) Series

(c) Cumulative compound

(d) Differential compound

Ans: **(a)**

Q. 103. However, if everything is kept unchanged, then the prime mover speed will have:

(a) Proportionality effect

(b) No effect

(c) Reciprocal effect

(d) May be either (a) or (c)

Ans: **(a)**

Q. 104. What is the general value of induced emf, due to residual magnetism?

(a) 10–25 V

(b) 30–35 V

(c) 5–15 V

(d) 17–35 V

Ans: **(c)**

Q. 105. Which one of the followings is the characteristic of a DC generator?

(a) No load and load saturation characteristic

(b) Internal or total characteristic

(c) External characteristic

(d) All of the above

Ans: **(d)**

Q. 106. The critical resistance in a DC generator is the resistance of:

(a) Load   (b) Brush

(c) Field   (d) Armature

Ans: **(c)**

Q. 107. The DC shunt generator shows:

    (a) Rising characteristics

    (b) Constant current characteristics

    (c) Continuously decreasing characteristics

    (d) Slightly drooping characteristics

Ans: **(d)**

Q. 108. In case of a self-excited DC generator, the magnetic characteristic starts from:

    (a) Below zero value of emf axis

    (b) Below zero value of current axis

    (c) Above zero value of emf axis

    (d) Above zero value on current axis

Ans: **(c)**

Q. 109. If a self-excited DC generator fails to build up during its first trial, after being installed, then the remedy lies in:

    (a) Reversing the field connections

    (b) Reducing the field and armature resistance

    (c) Increasing the prime mover speed

    (d) All of the above

Ans: **(a)**

Q. 110. A self-excited shunt generator fails to build up because:

    (a) Field or armature circuit is open

    (b) Residual magnetism is absent

    (c) The direction of rotation is not correct

    (d) All of the above

Ans: **(d)**

Q. 111. A direct current shunt generator, fails to build up voltage, it may be because:

    (a) There was a break in armature circuitry

    (b) Field resistance was connected in a wrong manner

(c) Field system lost residual magnetism

(d) Field resistance was higher than critical resistance

Ans: **(c)**

Q. 112. In a DC shunt generator, generator voltage may be increased by:

(a) Increasing the armature resistance

(b) Decreasing the armature resistance

(c) Increasing the speed or by increasing the field current

(d) Decreasing the speed

Ans: **(c)**

Q. 113. Some factors, which may affect the voltage building of a DC generator are:

(a) Reversed shunt field connection

(b) Reversed rotation

(c) Reversed residual magnetism

(d) All of the above

Ans: **(d)**

Q. 114. Whenever the shunt generator is loaded, the reasons for the drop in terminal voltage may be:

(a) Armature reaction drop

(b) Armature resistance drop

(c) Reduction in field current

(d) All of the above

Ans: **(d)**

Q. 115. Voltage regulation of a DC generator is the change in voltage:

(a) When load is reduced from rated value to 3/4th the rated value

(b) When load is increased from 0 to 1/4th the rated value

(c) When load is reduced from rated value to zero

(d) When load is increased from 0 to 1/2 the rated value

Ans: **(c)**

Q. 116. When the value of field resistance is increased beyond its critical value, then the generator:

(a) Will not be able to build up

(b) Power output will be less

(c) System will burn out

(d) Will bear heavy current through field winding

Ans: **(a)**

Q. 117. In an over compounded generator, the voltage regulation is always:

(a) High

(b) Low

(c) Positive

(d) Negative

Ans: **(d)**

Q. 118. Equalizing bus bar is necessary for the parallel operation. It may be used in:

(a) Series and over compounded generators

(b) Cumulative compound generators

(c) Shunt generators

(d) Series generators

Ans: **(a)**

Q. 119. Parallel operation of generators may be used, because:

(a) It provides continuity of service

(b) It provides high efficiency

(c) Both (a) and (b)

(d) None of the above

Ans: **(c)**

Q. 120. In which generator, equalizers are not required?

(a) Shunt generator

(b) Under-compounded generator

(c) Over-compounded generator

(d) Both (a) and (b)

Ans: **(d)**

Q. 121. Direct current generators are connected to or disconnected from the bus bar only under the floating condition. It may be done to avoid:

(a) Burning of switch contacts

(b) Sudden loading of prime mover

(c) Mechanical jerk to the shaft

(d) All of the above

Ans: **(d)**

Q. 122. When one of the generators is taken out from the service (In case of parallel operation of two generators), then:

(a) The excitation of second generator is reduced slowly

(b) The excitation of first is reduced and of the second is increased gradually and simultaneously

(c) The main switch of the first generator is short circuited

(d) The excitation of first generator is increased

Ans: **(b)**

Q. 123. The essential condition, which is needed for the parallel operation of two generators, is that they must have:

(a) Drooping voltage characteristics

(b) Same % impedance

(c) Same copper losses

(d) Different % impedance

Ans: **(a)**

Q. 124. In case of a generator, minimum amount of back torque will be present, when it is supplying:

(a) Half load current

(b) 25% of load current

(c) Zero ampere load current

(d) 2/3rd of load current

Ans: **(c)**

Q. 125. Maximum amount of back torque will be provided by DC shunt generator, when it supplies:

(a) 1/2 load current   (b) 1/4th load current

(c) Zero current   (d) Full load current

Ans: **(d)**

MISCELLANEOUS

Q. 1. In a direct current machine, the teeth in armature are sometimes skewed so as to reduce:

(a) Vibrations   (b) Copper losses

(c) Iron losses   (d) Eddy current losses

Ans: **(a)**

Q. 2. In DC machines, yoke or outer frame possesses:

(a) Sufficient mechanical strength

(b) High permeability

(c) Low permeability

(d) Both (a) and (b)

Ans: **(d)**

Q. 3. The pole pitch may be defined as:

(a) It is equal to the armature slots per pole

(b) It is the distance between two adjacent poles

(c) Either (a) or (b)

(d) None of the above

Ans: **(c)**

Q. 4. In direct current machines, which of the following parts are laminated:

(a) Armature and pole shoes

(b) Pole core

(c) Core

(d) Windings

Ans: **(a)**

Q. 5. In a wave wound DC machine, the number of parallel paths will be:

(a) $A = P$ (b) $A = 5$

(c) $A = 2P$ (d) $A = 2$

Ans: (d)

Q. 6. The ripples produced in the output of a DC generator may be minimized by using:

(a) Good quality brushes

(b) Copper as a winding material

(c) Very less commutator segments

(d) More commutator segments

Ans: (d)

Q. 7. In the pole shoes of DC machine, eddy currents are induced due to:

(a) Oscillating magnetic flux

(b) High resistance of path

(c) Low resistance of path

(d) Relative motion between field and armature

Ans: (d)

Q. 8. In a DC machine, induced emf in the coil will be maximum, when the plane of coil is:

(a) Parallel to the magnetic flux, during rotation

(b) Perpendicular to the magnetic flux, during rotation

(c) At $90° + \varphi$ to the magnetic flux, during rotation

(d) At any position, during rotation

Ans: (b)

Q. 9. In a DC machine, while using lap winding, the commutator pitch is always:

(a) ±3 or –3

(b) 4

(c) ±2 or –2

(d) Any value

Ans: (c)

Q. 10. The size of conductor in DC machine depends upon:

    (a) Rating of current

    (b) Conductivity and resistivity

    (c) KVA output and resistivity

    (d) Current rating and conductor resistivity

Ans: (d)

Q. 11. Soft carbon brushes are preferred in DC machine, because of:

    (a) Low value of brush contact drop

    (b) Low friction losses

    (c) Less damage to commutator

    (d) All of the above

Ans: (d)

Q. 12. In a DC machine, armature reaction is basically:

    (a) Active armature power

    (b) Apparent armature power

    (c) Magnetic effect of current flow in armature

    (d) All of the above

Ans: (c)

Q. 13. In a DC machine, maximum efficiency occurs, when:

    (a) Core becomes unsaturated

    (b) Core is fully saturated

    (c) Constant losses > Variable losses

    (d) Constant losses = Losses proportional to square of current

Ans: (d)

Q. 14. What is the shape of armature reaction MMF wave in DC machine?

    (a) Rectangular

    (b) Square

    (c) Sinusoidal

    (d) Traingular

Ans: (d)

Q. 15. The effect of armature reaction may be:

(a) Only demagnetizing

(b) Only cross magnetizing

(c) Both cross magnetizing and demagnetizing

(d) None of the above

Ans: (c)

Q. 16. Stator pole shoes are laminated to reduce:

(a) Iron loss

(b) Eddy current loss

(c) Hysteresis loss

(d) None of the above

Ans: (b)

Q. 17. In a direct current generator, the voltage due to residual magnetism is measured by running the armature, when the:

(a) Field is connected to load

(b) Field is not connected to load

(c) Field is connected to armature

(d) Field is not connected to armature

Ans: (d)

Q. 18. In case of a simplex lap winding, the conductors per pole must be:

(a) Always even   (b) Always odd

(c) May be even or odd   (d) None of the above

Ans: (c)

Q. 19. In lap winding, the values of back pitch and front pitch are always:

(a) Odd values

(b) Even values

(c) May be even or odd number

(d) None of the above

Ans: (a)

Q. 20. In a 6 pole DC generator, what will be the emf ratio, if they are wave wound and lap wound respectively?

(a) 3:2   (b) 3:1

(c) 4:2   (d) 4:1

Ans: **(b)**

Q. 21. The voltage output of a DC generator may be increased by:

(a) Decreasing the value of field resistance

(b) Increasing the speed of rotation

(c) Decreasing the speed of rotation

(d) Increasing the value of armature resistance

Ans: **(b)**

Q. 22. Due to larger air gap in DC machine:

(a) Higher reluctance to main field flux will be developed

(b) Total flux will reduce

(c) Total flux will increase

(d) Both (a) and (b)

Ans: **(d)**

Q. 23. The dynamically induced emf depends upon:

(a) Length of conductor

(b) Motion of conductor w.r.t field

(c) Magnetic field strength

(d) All of the above

Ans: **(d)**

Q. 24. Which losses are constant in a DC machine and do not depend on load?

(a) Iron losses

(b) Copper losses

(c) Friction and windage losses

(d) Both (a) and (c)

Ans: **(d)**

Q. 25. The armature core of a DC machine gets overheated due to:

(a) Armature copper loss

(b) Iron loss

(c) Friction loss

(d) Both (a) and (b)

Ans: **(d)**

Q. 26. The excitation of a shunt generator may be increased by:

(a) Increasing shunt field resistance

(b) Decreasing shunt field resistance

(c) Increasing armature resistance

(d) Decreasing armature resistance

Ans: **(b)**

Q. 27. In DC electric arc welding, which generator is employed?

(a) Shunt generator

(b) Series and compound generators

(c) Compound and shunt generators

(d) Compound generator

Ans: **(d)**

Q. 28. In case of a level compounded generator, the series field amp-turns are:

(a) Placed on the compensating winding

(b) In the same direction of shunt field amp-turns

(c) At 45° electrical to shunt field amp-turns

(d) In the same direction of series field amp-turns

Ans: **(b)**

Q. 29. Open type slots are preferred in DC machine armature, because:

(a) It reduces the coil reactance emf and aids in commutation

(b) It increases the output at load terminals

(c) It increases the induced emf per coil

(d) It increases the load bearing capacity

Ans: **(a)**

Q. 30. A DC shunt generator develops 450 A at a voltage of 230 volts. What will be the voltage drop in armature, if the values of R_{sh} and R_a are 50 ohms and 0.025 ohms respectively?

(a) 11.39 volts   (b) 38.4 volts

(c) 32.6 volts   (d) 39.92 volts

Ans: **(a)**

Q. 31. In a DC machine, the armature MMF is:

(a) Stationary in space and has trapezoidal space distribution

(b) Rotating at ½ full load speed

(c) Of rotating nature and has trapezoidal space distribution

(d) Stationary in space and has triangular space distribution

Ans: **(d)**

Q. 32. A direct current generator has fixed rpm. The short circuit current will be:

(a) Less than the maximum load current that it can feed

(b) Less than 3/4th the maximum load current

(c) More compared to full load current

(d) None of the above

Ans: **(a)**

Q. 33. Which of the following conditions are required to be satisfied for building up of voltage in DC shunt generator?

(a) The field resistance must be less compared to critical resistance corresponding to the speed of the machine

(b) The field winding must be connected in such a way that the current in the field winding produces flux in the same direction of residual magnetism

(c) Residual magnetism must not be present in the field system

(d) All of the above

Ans: **(d)**

Q. 34. The effects of bad commutation process include:

(a) Lesser output of generator

(b) More temperature rise in windings

(c) Sparking at the commutator surface

(d) All of the above

Ans: **(d)**

Q. 35. In a DC machine, first armature is wound for 2 pole lap winding, but after some processing it is converted into 2 pole wave winding. What will be the new induced emf?

(a) Half of the previous value

(b) It will be 1/4th the previous value

(c) It will remain same

(d) None of the above

Ans: **(c)**

Q. 36. When the armature is wound for wave winding, then how many number of brushes will be used?

(a) 2

(b) 4

(c) Equal to number of poles

(d) 8

Ans: **(a)**

Q. 37. What is the basic reason of the failure of current, in reaching to its maximum value in opposite direction, while undergoing commutation process?

(a) Lesser amount of current

(b) More current

(c) The reactance voltage

(d) Poor quality brushes

Ans: **(c)**

Q. 38. In a DC generator, to bring the brushes at MNP, the brushes are given:

(a) Forward lead in the direction of rotation of armature

(b) Backward lead in the direction of rotation of armature

(c) Backward lead in the opposite direction of rotation of armature

(d) Forward lead in the opposite direction of rotation of armature

Ans: **(a)**

Q. 39. What will be the relation between mechanical and electrical degrees for a DC machine having P number of poles?

(a) $\theta_{mech} = P/6\ \theta_{elec}$ (b) $\theta_{elec} = P/2\ \theta_{mech}$

(c) $\theta_{mech} = 6/P\ \theta_{elec}$ (d) $\theta_{mech} = P/4\ \theta_{elec}$

Ans: **(b)**

Q. 40. In a DC machine, the armature MMF wave is of:

(a) Trapezoidal shape and does not depend on load

(b) Triangular shape and is independent of speed

(c) Square shape and depends on impedance of load circuit

(d) Rectangular shape and depends on load circuit

Ans: **(b)**

Q. 41. What will be the number of armature amp-turns per pole for compensating winding?

(a) $0.7 \times ZI/2P$ (b) $0.5 \times ZI/P$

(c) $2 \times ZI/8P$ (d) $7 \times ZI/2P$

Ans: **(a)**

Q. 42. What must be the value of demagnetizing amp-turns of a 4-pole lap wound generator having 720 turns, giving 50 A? Assume a brush lead of 10° mechanical:

(a) 2500 A-T/pole   (b) 270 A-T/ pole

(c) 250 A-T/ pole   (d) 2700 A-T/ pole

Ans: **(c)**

Q. 43. What must be the value of heating time constant for a machine, which attains a temperature rise of nearly 60% of its final temperature rise in only one hour?

(a) 0.629 h   (b) 1.0914 h

(c) 2.02 h   (d) 3.902 h

Ans: (b)

Q. 44. In case of a direct current machine, the armature MMF axis and field flux axis are respectively along:

(a) Interpolar axis and direct axis

(b) Direct axis and interpolar axis

(c) Indirect axis and direct axis

(d) None of the above

Ans: (a)

Q. 45. The potential of 220 volts is built up at no load condition and at rated speed by a short shunt compound wound DC generator. However, if it is allowed to operate as a differentially compounded generator, then how much voltage would be created?

(a) 220 V ± 50 V

(b) Only 220 V

(c) Less than 220 V

(d) 270 V

Ans: (b)

Q. 46. Which one of the following statements is wrong?

(a) The brush axis remains along direct axis

(b) Armature MMF is directed along the brush axis

(c) The brush axis is along the quadrature axis

(d) The brush axis is not along quadrature axis

Ans: (d)

Q. 47. In direct current machines, which one of the following statements is wrong?

(a) Reactance voltage is proportional to length of core

(b) Reactance voltage is proportional to I_a

(c) Reactance voltage is proportional to self flux of coil

(d) Interpolar MMF is greater than armature MMF in the interpolar zone

Ans: (c)

Q. 48. Tapering shape is used for interpoles in a DC machine. It is actually done so as to:

(a) Increase the saturation in the interpole

(b) Reduce the saturation in the interpole

(c) Minimize the effect of commutation

(d) None of the above

Ans: (b)

Q. 49. The no load saturation curve of DC generator gives relation between:

(a) E_0/I_L (b) V/I_f

(c) E/I_a (d) E_0/I_f

Ans: (d)

Q. 50. Whenever the voltage drops due to armature reaction for various loads are subtracted from E_0, we may get the:

(a) Value of E

(b) Value of emf actually induced in armature under loaded conditions

(c) (a) or (b), as both have same meaning

(d) None of the above

Ans: (c)

Q. 51. A separately excited DC generator feeds a DC shunt motor. When the load torque on the motor is halved approximately, then the:

(a) Armature current of both generator and motor is halved

(b) Armature current of motor is doubled

(c) Armature current of generator is doubled

(d) Field current is halved

Ans: (a)

Q. 52. What are the basic reasons of hot bearings?

    (a) Bearing may be in a very tight mode

    (b) Belt may be too tight

    (c) Lack of oil

    (d) All of the above

Ans: (d)

Q. 53. Which factor causes armature heating?

    (a) Unequal strength of magnetic poles used

    (b) Moisture content, which short circuits the armature

    (c) Eddy current production

    (d) All of the above

Ans: (d)

Q. 54. The essential condition for stable parallel operation of two DC generators having similar characteristics is that they should have:

    (a) Same kilowatt output ratings

    (b) Drooping voltage characteristics

    (c) Same percentage regulation

    (d) Same no load and full load speed

Ans: (b)

Q. 55. The process of current commutation in a DC machine is opposed by the:

    (a) Resistance of coil   (b) Inductance of coil

    (c) Reactance voltage   (d) Supplied voltage

Ans: (c)

Q. 56. Why the no load voltage in a DC generator builds up to a finite steady value?

    (a) It is due to armature resistance

    (b) It is due to magnetic effects

    (c) It is due to magnetic saturation

    (d) It is due to field resistance

Ans: (c)

Q. 57. In DC machine, pole shoes are made larger compared to its pole body because:

(a) It reduces eddy current losses in the pole shoe

(b) It reduces hysteresis losses in the pole shoe

(c) It gives a more nearly rectangular flux density wave

(d) All of the above

Ans: **(d)**

Q. 58. Slot wedges are made of:

(a) Fiber   (b) Cast iron

(c) Mild steel   (d) Asbestos

Ans: **(a)**

Q. 59. A DC generator is running at 400 rpm without residual magnetism. However, if speed is raised to 800 rpm, what will be the effect on induced emf?

(a) It will be doubled

(b) It will be halved

(c) It will be zero

(d) It will be 1/4th the maximum value

Ans: **(c)**

Q. 60. The poles, which are of small size and placed in between the main poles are called as:

(a) Series poles   (b) Shunt poles

(c) Compensating poles   (d) Interpoles

Ans: **(d)**

Q. 61. At full load condition, the height of pole is decided by:

(a) Electromotive force   (b) Magnetomotive force

(c) Maximum flux   (d) Residual flux

Ans: **(b)**

Q. 62. What will be the torque developed in a DC machine?

(a) $T = \dfrac{1}{2\pi} \phi\, Z\, I_a\, (P/A)$

(b) $T = K_a\, \phi\, I_a$

(c) $T = \dfrac{1}{2\pi} \phi \, I_a \, (P/A)$

(d) Either (a) or (b)

Ans: **(d)**

Q. 63. Field resistance is that resistance:

(a) Above which the generator fails to excite

(b) Which equals the slope of resistance line tangential to open circuit characteristic

(c) Which equals the slope of resistance line perpendicular to open circuit characteristic

(d) Both (a) and (b)

Ans: **(d)**

Q. 64. In a DC machine, for a particular value of pole pitch (τ_p), what will be the value of flux per pole?

(a) $\varphi = B_{av} \, \tau_p \, l$

(b) $\varphi = B_{av} \, \tau_p / l$

(c) $\varphi = \tau_p \, l / B_{av}$

(d) $\varphi = B_{av} \, l / \tau_p$

Ans: **(a)**

Q. 65. In which region, the magnetic circuit of a DC machine operates?

(a) In unsaturated region of magnetization curve

(b) In somewhat saturated region of magnetization curve

(c) In the breakdown region of magnetization curve

(d) None of the above

Ans: **(b)**

Q. 66. Which of the following windings will be useful to provide greater emf in a DC generator for fixed number of conductors and poles?

(a) Wave winding only

(b) Lap winding only

(c) Both lap and wave winding

(d) None of the above

Ans: **(a)**

Q. 67. Which of the statements is definitely wrong?

(a) In a series dynamo, the circuit resistance must be less compared to critical resistance

(b) In a shunt dynamo, there is a lower limit for the resistance of external load, below which the machine will fail to excite

(c) There must be some residual magnetism in the field magnet

(d) In a series dynamo, the circuit resistance must be greater than critical resistance

Ans: (d)

Q. 68. Which of the following generators may be employed for increasing the voltage across the feeder, carrying current supplied by some other source?

(a) Shunt wound DC generator

(b) Series generator

(c) Short shunt compound generator

(d) Long shunt compound wound generator

Ans: (b)

Q. 69. In case of DC series generators, the load resistance may be more than its critical resistance. It may be due to:

(a) Greasy commutator surface

(b) High resistance of load circuit

(c) Open circuited system

(d) All of the above

Ans: (d)

Q. 70. A booster, which is inserted in circuit to add a certain voltage, may also be called as:

(a) Shunt generator

(b) Differential compound generator

(c) Series generator

(d) None of the above

Ans: (c)

FILL IN THE BLANKS

1. A DC machine may be an electric .... or generator.

Ans: Motor

2. The machine, which may convert mechanical energy into electrical energy or vice versa, is known as ....

Ans: Dynamo

3. Whenever conductor cuts the magnetic field, ... is developed in armature conductors.

Ans: Dynamically induced emf

4. In a DC machine, the object of the field system is to develop a .... field.

Ans: Uniform magnetic

5. In smaller machines, yokes are made of ....

Ans: Cast iron

6. The permeability of cast steel is nearly .... of cast iron.

Ans: Two times

7. Yoke possesses sufficient mechanical strength and .... permeability.

Ans: High

8. The pole shoe acts as a support to .... coils

Ans: Field

9(a). Pole core is used to carry the coils of insulated wires carrying the .... current.

Ans: Field

9(b). The output voltage increases with increase in .... for a constant speed of armature.

Ans: Number of poles

9(c). Air gap in between the pole pieces and armature is kept ....

Ans: Small

9(d). The armature is supported at each end by a metal framework known as ....

Ans: End bells

9(e). Commutator is employed in a DC machine to provide electrical connection between the rotating armature coils and ....

Ans: Stationary external circuit

9(f). Brushes are staggered so as to prevent .... on the surface of commutator.

Ans: Ridge formation

9(g). Various rotating parts such as commutator, cooling fan, etc. are keyed to the ....

Ans: Shaft

9(h). The total number of conductors per pole may be known as .... pitch.

Ans: Pole

9(i). When the coil span of the winding is equal to ...., the coils are known as full pitched coils.

Ans: Pole pitch

9(j). Wave winding may also be called as ....

Ans: Series winding

10. The brushes are used in DC machine so as to collect .... from the commutator and supply it to ....

Ans: Current, external load circuit

11. The armature of a DC machine is made of laminated silicon steel so as to reduce ....

Ans: Iron losses

12. Air gap is kept very small so as to keep the .... of magnetic path very low.

Ans: Reluctance

13. The copper brushes are hard and may spoil the .... segments.

Ans: Commutator

14. The windings in a DC machine are placed in ....

Ans: Slots

15. The effect of magnetic field produced by the armature current on the distribution of flux under the main poles of DC machine is called as ....

Ans: Armature reaction

16. MNA is known as ....

Ans: Magnetic neutral axis

17. .... is generally preferred for the yoke in a DC machine.

Ans: Cast steel

18(a). In order to have satisfactory commutation process, the coils short circuited by the brushes, must possess .... induced in them.

Ans: Zero emf

18(b). When the brushes of DC machine lie along geometrical neutral axis, the armature reaction effect will be totally ....

Ans: Cross magnetizing

18(c). In commutation process, alternating current changes to an externally available ....

Ans: Direct current

19. In good commutation process, there will be no .... at the brushes.

Ans: Sparking

20. Self induced emf is also known as .... in the coil undergoing commutation.

Ans: Reactance voltage

21. Reactance voltage is proportional to the square of the number of .... in the armature coil.

Ans: Turns

22. By using interpoles, sparkless commutation may be achieved up to .... overload.

Ans: Nearly 20% to 30%

23. For a given value of output, an interpolar machine is made .... compared to a non-interpolar machine.

Ans: **Smaller and cheaper**

24. To provide uniform flux distribution under the faces of main poles .... winding may be employed.

Ans: **Compensating**

25. Armature reaction is more severe in .... machines.

Ans: **Large size and high speed**

26. Compensating winding is actually an .... winding, which is embedded in slots located in the faces of main poles.

Ans: **Auxiliary**

27. The gaps, under the interpoles are kept longer than that under the main poles, so as to .... interpoles.

Ans: **Avoid saturation in**

28. A permanent magnet generator consists of an armature and several permanent magnets encircling the ....

Ans: **Armature**

29. Permanent magnet generators are used in small sizes in ...., etc.

Ans: **Motor cycles**

30. In a separately excited DC generator, field coils are supplied by a .... source.

Ans: **DC**

31. In a separately excited DC generator, power developed is the product of load current and ....

Ans: **Generated emf**

32. Some amount of flux always remains in the system due to..........

Ans: **Residual magnetism**

33. In a series wound generator, there is only one .... which is connected in series with armature.

Ans: **Field winding**

34. In series generator, power delivered to the load is ....

Ans: **VI**

35. The shunt field winding resistance in shunt generator is of the order of .... ohm.

Ans: **100**

36. In a compound generator, the major portion of excitation is supplied by .... field.

Ans: **Shunt**

37. Compound generators are of two types, namely cumulative compound and .... generators.

Ans: **Differential compound wound**

38. The value of R_a mainly depends upon .... type.

Ans: **Machine**

39. The value of emf generated is directly proportional to the product of flux per pole and ....

Ans: **Speed**

40. The no load saturation characteristic of DC generator is also called as ....

Ans: **Open circuit characteristic**

41. Total characteristic may also be known as .... characteristic.

Ans: **Internal**

42. The external characteristic gives the relation between ....

Ans: **Terminal voltage and load current**

43. The separately excited DC generator has the advantage that it operates in a .... condition, with any value of field excitation.

Ans: **Stable**

44. The critical resistance of generator is equal to the .... line.

Ans: **Slope of tangent**

45. In a cumulative compound wound DC generator, an additional winding, having few turns of thick wire, in

series with either line or armature is used, called as
.... winding.

Ans: Series field

46. In a flat compounded generator, the terminal voltage is not constant for all loads i.e. from .... load to .... load.

Ans: No; Full

47. Whenever in DC machines, the series excitation becomes more prominent compared to shunt field, then the terminal voltage rises with increase in load and generator is called as .... generator.

Ans: Over compounded

48. When shunt field excitation becomes predominant, terminal voltage becomes less compared to no load terminal voltage, then the generator is known as .... generator.

Ans: Under compounded

49. In a cumulative compound wound generator, characteristics of the .... and .... generators are combined.

Ans: Shunt; Series

50. A diverter is basically a .... shunt.

Ans: Low resistance

51. The field current circuit must never be broken except by means of ....

Ans: Special switches

52. In a series generator, the resistance of external circuit must be .... compared to critical resistance.

Ans: Less

53. In a shunt generator, the resistance of shunt field circuit may be .... than critical resistance.

Ans: Greater

54. Shunt generators may be used for.......... purposes with field regulators.

Ans: Light and power supply

55. Critical resistance is the .... value of resistance with which generator will be able to excite.

Ans: **Maximum (in series generators)**

56. Parallel operation of generators keeps the machine loaded up to their .... capacity.

Ans: **Rated**

57. To have satisfactory parallel operation, it becomes necessary to connect the armatures of the two machines by a heavy copper bar, known as ....

Ans: **Equalizing bar**

58. For parallel operation of DC series generators, field winding may be connected as .... arrangement.

Ans: **Cross connection**

59. Due to the ....characteristic, shunt generators are best suited for parallel operation.

Ans: **Drooping**

60. The generator, which has more drooping characteristic, carries .... load.

Ans: **Small**

61. For parallel operation, the polarity of the incoming generator must be .... as that of the generator, previously installed.

Ans: **Same**

62. Whenever one of the generators is taken out of service, its main switch must never be .... suddenly.

Ans: **Opened**

63. The generators, operating in parallel must have the same voltage regulation so as to share the total load in proportion to their ....

Ans: **Respective rated capacity**

64. In under compounded generators, equalizing bus bars are not needed for .... operation.

Ans: **Parallel**

65. In normal DC machines, stator core is ....

Ans: **Not laminated**

66. Direct current machines, used in control system have their magnetic circuit completely ………

Ans: Laminated

67. In a DC machine, armature core handles nearly ………. portion of flux.

Ans: Half

68. The flux carried by yoke is stationary, so it is ……….

Ans: Not laminated

69. Field pole consists of ………. core and ……….

Ans: Pole; Pole shoes

70. In a DC series machine ………. number of turns of greater cross-section are employed.

Ans: Smaller

71. The interpoles are normally tapered with sufficient sectional area at the root so as to avoid ……….

Ans: Magnetic saturation

72. Compensating winding is generally connected in series with ………. circuit.

Ans: Armature

73. The windings are of two types namely ……….

Ans: Lap and wave

74. Commutator is of ………. structure.

Ans: Cylindrical

75. A DC machine can work as an ………. energy converter.

Ans: Electro–mechanical

76. Series field current depends on the armature current and so called as ………. operated field.

Ans: Current

77. A shunt field may also be called as ………. operated field.

Ans: Voltage

78. In a compound excitation, both the series excited winding and a ………. may be used.

Ans: Shunt excited winding

79. When the flux produced by a series field helps the shunt field flux, then the resultant air gap flux per pole is ....

Ans: **Increased**

TRUE / FALSE

1. In generator, energy conversion is based on the principle of the production of dynamically induced emf.

Ans: **True**

2. For smaller DC machines, cast steel may be used.

Ans: **False**

3. Armature winding is that winding, which produces the working flux.

Ans: **False**

4. The current in a winding, which varies as the machine is loaded is called load current.

Ans: **True**

5. In DC machines, field magnets consist of only pole cores.

Ans: **False**

6. The purpose of pole core is to protect exciting coils.

Ans: **False**

7. In a DC machine, the pole core and pole shoes are built of thin laminations.

Ans: **True**

8. The field coils are also called as pole coils.

Ans: **True**

9. The purpose of armature core is to provide a path of very low reluctance to the flux, through the armature from a N-pole to S-pole.

Ans: **True**

10. The laminations are used so as to reduce eddy current losses.

Ans: **True**

11. Commutator is of a cylindrical structure.

Ans: **True**

12. Every commutator segment is connected to the armature conductor by using copper lugs or stripes.

Ans: **True**

13. Pole cores are generally not laminated and made of cast steel.

Ans: **True**

14. The pole faces or pole shoes are always laminated to avoid heating and eddy current losses caused by the fluctuation.

Ans: **True**

15. The yoke carries the flux from pole to pole.

Ans: **True**

16. The armature is the stationary part of DC machine.

Ans: **False**

17. The flux density in the body of the pole core is much lesser compared to that permitted in air gap.

Ans: **False**

18. Core punching up to a diameter of 70 cm are often made in only one piece.

Ans: **False**

19. The connection between the armature and external circuit is made through the brushes.

Ans: **True**

20. The individual brushes are supported in plastic holders.

Ans: **False**

21. The DC machine windings are only of drum type.

Ans: **False**

22. Simplex wave windings do not need equalizer rings.

Ans: **True**

23. A wave winding does not usually require equalizer rings.

Ans: **True**

24. Lap windings may be employed in high power machines so as to reduce current per armature path.

Ans: **True**

25. The periphery of the armature divided by the number of poles is known as coil span.

Ans: **False**

26. Coil span means coil pitch.

Ans: **True**

27. The distance, which is measured in terms of armature conductors, which a coil advances on the back of the armature is known as back pitch.

Ans: **True**

28. In single layer windings, one conductor is placed in each armature slot.

Ans: **True**

29. Hysteresis loss is due to the reversal of magnetization of armature core.

Ans: **True**

30. Hysteresis loss depends only upon the grade of iron.

Ans: **False**

31. Eddy current loss depends only upon thickness of lamination.

Ans: **False**

32. In a DC generator, copper losses are nearly 30% to 40% of the full load losses.

Ans: **True**

33. Mechanical losses include friction and windage losses in a DC machine.

Ans: **True**

34. Stray losses and shunt copper losses are considered as the variable losses.

Ans: **False**

35. The generator gives maximum efficiency, when variable loss equals constant loss.

Ans: **True**

36. The DC armature winding, in which coil sides are a pole pitch apart is called short pitch winding.

Ans: **False**

37. In a DC generator, the generated value of emf is directly proportional to the pole flux.

Ans: **True**

38. The demagnetizing effect of armature reaction reduces generated voltage.

Ans: **True**

39. The cross magnetizing effect of armature reaction produces sparking at the brushes.

Ans: **True**

40. Magnetic neutral axis is that axis along which no emf is produced in the armature conductors.

Ans: **True**

41. Both the distorting as well as demagnetizing effects of armature reaction will decrease with increase in armature current.

Ans: **False**

42. The function of compensating winding is to neutralize the demagnetizing effect of armature reaction.

Ans: **False**

43. Reactance voltage may also be called as self-induced emf.

Ans: **True**

44. Reactance voltage is the ratio of self inductance to rate of change in voltage.

Ans: **False**

45. Methods of improving commutation may be resistance commutation and emf commutation.

Ans: **True**

46. The main reason of sparking commutation is the self-induced emf.

Ans: True

47. The commutator is made larger as compared to the cu brushes.

Ans: True

48. Interpole may also be known as compensating winding.

Ans: False

49. An interpolar machine is made larger and costlier compared to a non-interpolar machine.

Ans: False

50. In lap winding, all conductors in any parallel path lie under one pair of poles.

Ans: True

51. The equalizing conductors are called as annular rings.

Ans: False

52. Equalizer rings are often used in wave wound armatures.

Ans: False

53. When all the conductors are connected to an equalizer ring, then the winding is called as 57.7% equalized.

Ans: False

54. The basic function of equalizer rings is to avoid unequal distribution of current at the brushes, thus helping to get sparkless commutation.

Ans: True

55. Various DC generators may be operated in parallel mode.

Ans: True

56. Generators operate most efficiently, when delivering full load.

Ans: True

57. Due to slightly drooping voltage characteristics, compound generators are most suited for stable parallel operation.

Ans: False

58. Shunt and under compounded generators need equalizer rings for satisfactory parallel operation.

Ans: **False**

59. The main factor, which leads to unstable parallel operation of flat and over compounded DC generators is their unequal series field resistance.

Ans: **False**

60. Performance characteristic may also be known as voltage regulating curve.

Ans: **True**

61. Total characteristic of a DC generator may be drawn between E and I_a.

Ans: **True**

62. OCC is also known as magnetic characteristic of DC generator.

Ans: **True**

63. The resistance, represented by the tangent to the curve is called as critical resistance.

Ans: **True**

64. Reversed residual magnetism, affect the voltage building of a self-excited DC generator.

Ans: **True**

65. After building up, a shunt generator is loaded and its terminal voltage (V) decreases with decrease in load current.

Ans: **False**

66. Armature resistance and armature reaction are only the responsible factors for the drop in terminal voltage in a DC shunt generator.

Ans: **False**

67. Voltage regulation of generator means the change in voltage from rated value to ½ the rated value.

Ans: **False**

68. A booster may also be called as series generator.

Ans: **True**

69. When the series field amp-turns in a compound generator are in such a manner that rated load voltage is greater than the no load voltage, then the generator is known as flat compounded generator.

Ans: False

70. Differential compounded generator may not be used in arc welding process.

Ans: False

71. In a DC generator, the voltage building process is cumulative.

Ans: True

72. In an ideal DC generator, voltage regulation is negative.

Ans: False

73. In a well designed DC machine, the permissible rise in temperature may be 57 °C.

Ans: False

74. In a DC generator, efficiency in the range of 90% to 95% may be obtained.

Ans: False

75. In DC generators, compound generators may be classified as flat-compounded, over-compounded and under-compounded.

Ans: True

VIVA VOCE QUESTIONS

Q. 1. What do you mean by a DC generator?

Ans: It is a machine that converts mechanical energy into electrical energy.

Q. 2. On which principle, the energy conversion is based?

Ans: It is based on the principle of the production of dynamically induced emf.

Q. 3. What are the basic essential parts of a DC generator?

Ans: The parts include the following:

(a) Yoke (outer frame)

(b) Pole core and pole shoes

(c) Field coils and armature core

(d) Armature windings

(e) Commutator

(f) Brushes and bearings

Q. 4. What is the purpose of yoke?

Ans: It provides:

(a) Support to the poles and acts as a protecting cover for the whole machine and

(b) Carries the magnetic flux produced by poles.

Q. 5. What are the two main components of field magnet?

Ans: It actually has pole core and pole shoes.

Q. 6. What is the function of pole shoes in DC machines?

Ans: The functions include:

(a) It spreads out the flux over the whole periphery of armature in a more uniform manner. Since it has a large cross-sectional area, so due to it $\left(\because \ S = \dfrac{l}{\mu A} \right)$ the reluctance of magnetic path is reduced.

(b) It also supports the field coils (exciting coils).

Q. 7. What is the purpose of armature core?

Ans: The armature core provides space for armature coils and so causes them to rotate. In addition to this, it also provides a path of very low reluctance to flux through the armature from N-pole to S-pole.

Q. 8. What is the function of commutator?

Ans: Its main function is to change induced AC emf into DC for external circuit.

Q. 9. What is the function of brushes?

Ans: Its basic function is to collect current from commutator and to deliver for external load.

Q. 10. What capabilities must be possessed by the yoke?

Ans: The yoke must possess sufficient mechanical strength and high permeability.

Q. 11. Why the field coils are provided in DC machines?

Ans: These coils are provided so as to give required number of ampere turns of excitation, needed to give proper flux through the armature, so that required emf may be produced.

Q. 12. What may be the disadvantages of using more number of poles?

Ans: The disadvantages may be:
(a) Increase in labour charges.
(b) Increased tendency of flash over between brush arms.
(c) Iron losses may get increased.

Q. 13. What is the necessity to keep air gap between pole pieces and armature?

Ans: It is needed so as to avoid rubbing in the machine.

Q. 14. For which machines, copper brushes are employed?

Ans: These brushes are used for machines designed for large currents at low voltage.

Q. 15. What is the desired pressure, applied on the brushes in DC machine?

Ans: The required pressure on the brushes is nearly in the range of 1.5–2.5 N/cm^2 to provide satisfactory commutation.

Q. 16. Which material is used for shaft?

Ans: It is made of mild steel so as to give good breaking strength.

Q. 17. What do you mean by the term winding?

Ans: Number of coils arranged in coil groups is called winding.

Q. 18. What is the meaning of pole pitch?

Ans: It is actually the number of conductors per pole.

Q. 19. Why the coils have various turns in series in commercial generators?

Ans: The main purpose is to increase the magnitude of generated emf, being in direct proportion to the number of turns in the coil.

Q. 20. What is the purpose of laminating the armature core of a DC generator?

Ans: The purpose is to reduce the eddy current losses.

Q. 21. In which type of machine, a dummy coil is used in armature winding?

Ans: In wave wound machine.

Q. 22. What is the advantage of making armature with laminations of silicon steel?

Ans: It is used to reduce the eddy current and hysteresis losses to minimum.

Q. 23. What provisions are made for cooling purpose in DC machines?

Ans: In these DC machines, horizontal ventilating ducts and fans are provided for the purpose of cooling.

Q. 24. What is the basic working principle of direct current generator?

Ans: Whenever a conductor cuts the magnetic flux, an emf is induced in it due to the electromagnetic induction phenomenon called as dynamically induced emf and this induced emf is proportional to the rate of change of flux.

Q. 25. What do you mean by armature reaction?

Ans: The effect of magnetic field, set up by the armature current due to the distribution of flux under the main poles of a DC machine is known as "armature reaction".

Q. 26. What do you mean by magnetic neutral plane?

Ans: The plane through the axis, along which no emf is induced in the armature conductors is known as magnetic neutral plane.

Q. 27. What are the bad effects of armature reaction?

Ans: The following are the bad effects:
  (a) Weakens and distorts the main magnetic field
  (b) Reduces the value of induced voltage
  (c) Produces sparking on the surface of commutator
  (d) Changes the MNP position

Q. 28. What do you mean by good commutation?

Ans: Good commutation means–there will be no sparking at the brushes and the surface of commutator will remain unaffected during the operation of DC machine.

Q. 29. What is the meaning of poor commutation?

Ans: It simply means that sparking will definitely take place at brushes and the commutator surface will get damaged during the operation of DC machine.

Q. 30. What is the essential condition for satisfactory commutation?

Ans: The essential condition is that the current in a coil undergoing commutation must be completely reversed during the commutation period.

Q. 31. Due to which reason, delay in reversal of current takes place?

Ans: It is due to the inductive property.

Q. 32. What is the basic meaning of commutation?

Ans: It means the changes taking place in an armature coil during the period of short circuit by the brushes, when it moves toward opposite poles.

Q. 33. What is the expression for reactance voltage?

Ans: It can be expressed as
  $V_R = L \times 2I/t$, where L = inductance in henry
  I = current in the coil
  t = time of commutation

Q. 34. On which factors, time of commutation depends?

Ans: The time of commutation, is that time, which the commutator needs in moving a distance equal to circumferential thickness of brush minus the thickness of mica insulation between two commutator segments, i.e.

$$t_c = \frac{(W_b - W_m)}{V_c} \text{ seconds}$$

Q. 35. What are the disadvantages of commutation process?

Ans: The disadvantages are as follows:
(a) Lesser output of generator
(b) More temperature rise in armature winding
(c) Due to sparking at commutator, its life reduces

Q. 36. By which process, commutation can be improved?

Ans: It may be improved by the following processes:
(a) Use of high resistance brushes
(b) Shift of brushes
(c) Use of interpoles or commutating poles

Q. 37. Why the commutating poles are used in large DC machines?

Ans: In case of large DC machines, armature reaction effect is very severe and the commutation problem increases due to large armature currents and to avoid all these problems, commutating poles are used.

Q. 38. What is the basic meaning of compensating winding?

Ans: It is an auxiliary winding, which is used to neutralize the cross magnetizing effect of armature reaction. These windings are embedded in slots in the pole shoes. Actually these are employed for large sized direct current machines, which are subjected to large fluctuations in load.

Q. 39. What do we mean by interpoles?

Ans: These are small poles fixed to yoke and placed in between the main poles. The function of interpole is as follows:

(a) The emf produced by interpoles (commutating emf) neutralizes the reactance emf, thus gives sparkless commutation.

(b) It also neutralizes the cross magnetizing effect of armature reaction.

Q. 40. What are slots?

Ans: Slots are basically grooves, which provide the location for conductors. These may be open type; semi enclosed type or totally closed type.

Q. 41. What is the type of DC armature winding?

Ans: DC armature windings are closed type windings.

Q. 42. What are the two main types of DC winding?

Ans: It includes:
(a) Lap winding and
(b) Wave winding

In lap winding, the number of parallel paths are same as that of the number of poles in the machine. Likewise, wave winding is that one, which has only two parallel paths, whatsoever be the number of poles.

Q. 43. Which generators are of electromagnet type?

Ans: These (electromagnet type) generators may be
(a) Separately excited generators and
(b) Self-excited generators

Q. 44. Why the emf is induced in self-excited generators?

Ans: It is due to residual magnetism.

Q. 45. What are the basic types of self-excited generator?

Ans: It may include:
(a) Series generator
(b) Shunt generator
(c) Compound wound generator

Q. 46. What are the two basic types of compound wound generator?

Ans: It consists of two types namely:
(a) Short shunt compound wound generator and
(b) Long shunt compound wound generator

Q. 47. What do we mean by short shunt and long shunt compound wound generators?

Ans: Short shunt generator is that generator, in which the shunt field winding is connected in parallel across the armature terminals. Likewise, in long shunt generator, shunt field winding is connected in parallel to the series combination of series winding and armature.

Q. 48. In which generator, field current is same as that of load current?

Ans: Series generator

Q. 49. In which generator, both series and shunt field windings are used?

Ans: Compound wound DC generator

Q. 50. In which generator, shunt field winding is connected in parallel to armature terminals?

Ans: DC shunt wound generator

Q. 51. What do you mean by critical resistance of DC series generator?

Ans: It is the maximum value of resistance, with which DC series generator will be able to excite.

Q. 52. What are the advantages of parallel operation?

Ans: The advantages are as follows:
(a) Maintenance and repair are easy for generators connected in parallel mode.
(b) Continuity of service is maintained.
(c) High efficiency.
(d) Facility for additions to power plant.

Q. 53. What is the actual requirement of parallel operation?

Ans: It is required for the following purposes:
  (a) To meet the actual load demand
  (b) To provide continuous supply in case of failure of any generator
  (c) To provide maximum efficiency of the system

Q. 54. What are the basic conditions, which are required for satisfactory parallel operation of DC generators?

Ans: The conditions are:
  (a) Same polarity and   (b) Equal terminal voltage

Q. 55. How the parallel operation of two DC series generators is made stable?

Ans: By connecting the armatures of two machines (operating in parallel) through a heavy copper bar or by cross connecting their field windings, the parallel operation may be made stable.

Q. 56. Why the two generators should have the same voltage regulation, while operating in parallel mode?

Ans: To share the load, in proportion to their rated capacity, the two generators should have the same voltage regulation.

Q. 57. How the load shifting may take place between generators?

Ans: By increasing the field excitation of the generator to be loaded.

Q. 58. What is equalizing bus bar?

Ans: It is that bus bar, which is used to share the load on DC generators. It actually helps to load a generator and provides automatic distribution of load in running condition.

Q. 59. What is the meaning of bus bars?

Ans: Whenever a number of generators operate in parallel mode, positive and negative terminals of all the machines are connected to two separate heavy copper bars, placed behind the switch panel. These copper bars are called as bus bars.

Q. 60. Why a DC generator gets overheated?

Ans: It may be due to following reasons:

(a) Defect in bearings used.

(b) Sparking on the commutator surface.

(c) Short circuiting in the field coil or armature.

2

Direct Current Motor

MULTIPLE CHOICE QUESTIONS

Q. 1. Which effect is produced by electric current in DC motor?

(a) Chemical and heating effects

(b) Chemical effect only

(c) Magnetic and heating effects

(d) None of the above

Ans: (c)

Q. 2. Which type of motor is most suitable for traction purpose?

(a) Differential compound motor

(b) Shunt motor

(c) Series motor

(d) All of the above

Ans: (c)

Q. 3. When a DC motor is switched on/started, then it rotates in the direction opposite to that for which it is designed. What is this type of motor called?

(a) Either series or shunt wound motor

(b) Differential compound motor

(c) Series motor

(d) Shunt motor

Ans: (b)

Q. 4. In a DC motor, the emf induced is generally known as:

(a) Counter emf

(b) Opposing emf or back emf

(c) Generated emf

(d) Both (a) and (b)

Ans: **(d)**

Q. 5: On which factor, counter emf in a DC motor depends?

(a) Flux per pole

(b) Speed of rotation of armature

(c) Number of poles

(d) All of the above

Ans: **(d)**

Q. 6. In DC motor, the gross mechanical power developed is the maximum, when:

(a) Counter emf is greater than applied voltage

(b) Back emf is 1/4th of the applied voltage

(c) Back emf is equal to half the applied voltage

(d) None of the above

Ans: **(c)**

Q. 7. In case of DC motor, armature torque developed does not depend upon:

(a) Total number of poles (b) Armature current

(c) Constant losses (d) Back emf

Ans: **(c)**

Q. 8. The speed of a DC motor depends upon:

(a) Flux (b) Back emf

(c) Both (a) and (b) (d) Armature current only

Ans: **(c)**

Q. 9. The speed regulation of DC motor may be defined as:

(a) The change in speed, when the load is reduced from rated value to zero, expressed as percent of the rated load speed

(b) The change in speed, when the load is reduced from rated value to half the rated value

(c) The change in speed, when the load is increased from rated value to twice the rated value

(d) None of the above

Ans: (a)

Q. 10. In DC shunt motors, the excitation of field is kept at the maximum value during the starting period. It is done so as to:

(a) Reduce the heating of armature

(b) Increase the armature current

(c) Reduce voltage dips in the supply mains

(d) Reduce the armature current

Ans: (a)

Q. 11. The speed of a DC motor is inversely proportional to:

(a) Flux   (b) Back emf

(c) Both (a) and (b)   (d) None

Ans: (a)

Q. 12. Which of these characteristics is named as mechanical characteristic of a DC motor?

(a) Speed and armature current (N/I_a)

(b) Speed and torque (N/T_a)

(c) Torque and armature current (T_a/I_a)

(d) None of the above

Ans: (b)

Q. 13. A DC series motor is basically:

(a) Variable speed motor

(b) Constant speed motor

(c) Adjustable speed motor

(d) Both (b) and (c)

Ans: (a)

Q. 14. Up to the point of magnetic saturation, T_a/I_a curve of a DC series motor is of:

(a) Parabolic shape

(b) Hyperbolic shape

(c) Linear shape

(d) None of the above

Ans: **(a)**

Q. 15. At places, where highest starting torque is needed, the motor to be used is:

(a) AC series motor

(b) DC shunt motor

(c) Cumulative compound motor

(d) DC series motor

Ans: **(d)**

Q. 16. DC shunt motor is taken as:

(a) Constant speed motor   (b) Adjustable speed motor

(c) Neither (a) nor (b)   (d) Variable speed motor

Ans: **(a)**

Q. 17. In a DC shunt motor, torque developed is directly proportional to:

(a) Armature current

(b) Field current

(c) Square of armature current

(d) Back emf

Ans: **(a)**

Q. 18. In a DC series motor, if the armature current is reduced by 50%, the torque of motor will be equal to:

(a) 50% of previous value

(b) 25% of previous value

(c) 27% of previous value

(d) 80% of previous value

Ans: **(b)**

Q. 19. Whenever the brushes are given a shift from the interpolar axis in the direction of rotation in a loaded DC motor, the commutation process will:

(a) Deteriorate and the speed falls

(b) Improve

(c) Be called as good commutation

(d) None of the above

Ans: **(a)**

Q. 20. Which one of the following statements is correct?

(a) Interpoles have their windings connected in series with the armature

(b) The purpose of interpoles is to reduce armature reaction effects in the interpolar region

(c) Both (a) and (b)

(d) Interpoles have their windings connected in series with field

Ans: **(c)**

Q. 21. The interpoles have tapering shape in a DC machine. It is done so as to reduce:

(a) Overall size of the complete system

(b) Sparking at the brushes

(c) The saturation in the interpole

(d) Reactance voltage produced in the coil

Ans: **(c)**

Q. 22. Sparking takes place at the commutator of a normal DC motor, because of the:

(a) Reactance voltage

(b) Field winding current

(c) Eddy current produced

(d) Low value of impressed voltage

Ans: **(a)**

Q. 23. The method of neutralizing the effect of armature reaction for better commutation includes:

(a) Use of high brush-contact resistance

(b) Shifting of brushes

(c) Use of commutating poles and compensating winding

(d) All of the above

Ans: **(d)**

Q. 24. The losses in case of a DC machine include:

(a) I^2R losses in the armature winding

(b) Iron losses in the armature core

(c) Field winding and frictional losses

(d) All of the above

Ans: **(d)**

Q. 25. The eddy current loss in a DC machine depends upon:

(a) The maximum value of flux density

(b) Core material and thickness of core laminations

(c) Frequency of magnetic reversals

(d) All of the above

Ans: **(d)**

Q. 26. The friction losses in a DC machine include:

(a) Brush friction loss   (b) Windage friction loss

(c) Bearing friction loss   (d) All of the above

Ans: **(d)**

Q. 27. In DC machines, constant loss is composed of:

(a) Iron loss and mechanical loss

(b) Friction and windage losses (FW), iron loss and field circuit loss

(c) Stray load loss and iron loss

(d) Magnetic loss

Ans: **(b)**

Q. 28. Two DC shunt machines are coupled mechanically and electrically. One of them runs as a motor and other as a generator. Their no load rotational losses are equal, because:

(a) Stray losses are more in motor as compared to generator

(b) Motor iron losses are less as compared to generator iron losses

(c) Both the machines have equal speed

(d) All of the above

Ans. **(d)**

Q. 29. In a DC compound motor, the armature circuit ohmic losses include:

(a) Shunt I^2R loss

(b) Brush contact loss and series field loss

(c) Series field I^2R loss, brush contact loss and armature I^2R loss

(d) Series and shunt I^2R losses

Ans: (c)

Q. 30. The three point starter, is used in DC machines so as to start:

(a) DC series motor   (b) DC shunt motor

(c) Universal motor   (d) AC series motor

Ans: (b)

Q. 31. While using a DC three point starter, the overload tripping contact actually short circuits the:

(a) No volt release coil

(b) Armature winding

(c) Field and armature windings

(d) Both (a) and (c)

Ans: (a)

Q. 32. What will happen, if a DC shunt motor is started with open circuited field winding?

(a) The motor does not pick up speed and draws large current

(b) The motor picks up speed (rated speed)

(c) The motor draws small current and runs at fast speed

(d) The motor does not pick up speed and draws small current

Ans: (a)

Q. 33. Whenever a DC shunt motor is started, the:

(a) Rated armature and rated field voltage must be applied

(b) Reduced armature voltage and full field voltage should be applied

(c) Whole regulator resistance must be cut out in the field winding circuit

(d) Both (b) and (c)

Ans: **(d)**

Q. 34. The role of starter in a DC motor is to:

(a) Reduce sparking at the brushes

(b) Reduce back emf

(c) Limit the starting current to a safer value

(d) Provide overload protection

Ans: **(c)**

Q. 35. Which one of the following statements is incorrect?

(a) In a DC compound motor, 4 point starter may be used to provide under voltage and over load protection

(b) A 3 point DC starter may be used for DC shunt motor

(c) In a DC motor, the starter-handle must be moved very slowly in steps

(d) In a DC compound wound motor, 4 point starter is used to provide wide range of speed control

Ans: **(c)**

Q. 36. At starting moment of DC motor, the DC motor draws:

(a) Zero current

(b) High value of current

(c) Very low current

(d) 5 times the rated current

Ans: **(c)**

Q. 37. In a DC motor, to bring the handle back to "OFF" position, the portion is used with the handle, known as:

(a) Soft iron segment   (b) Strip

(c) Spring   (d) Tripping contact

Ans: **(c)**

Q. 38. In a DC motor (provided with a 4 point starter), the no volt release coil circuit is:

(a) Connected in parallel with a resistance of very high value

(b) Connected in series with a low value of resistance

(c) Made independent

(d) Open circuited for the starting instant

Ans: (c)

Q. 39. While starting a differentially connected compound motor, it is required to short circuit the series field winding to:

(a) Provide safety

(b) Reduce the armature reaction MMF

(c) Both (a) and (b)

(d) Avoid excessive starting speed

Ans: (d)

Q. 40. When a DC shunt motor is started, then:

(a) Rated armature voltage and reduced field voltage must be applied

(b) Rated field voltage and rated armature voltage must be applied

(c) Reduced armature voltage and full field voltage must be applied

(d) None of the above

Ans: (c)

Q. 41. The purpose of keeping the field excitation of a DC shunt motor at its maximum value (at starting) is to reduce:

(a) Sparking at brushes

(b) The value of starting current

(c) Voltage dips in supply

(d) All of the above

Ans: (d)

Q. 42. **Which one of the following statements is not correct regarding a DC shunt motor starter?**

(a) It reduces the acceleration time

(b) It provides starting torque, which is always more compared to load torque

(c) It maintains the field flux at the maximum value

(d) It avoids the field failure

Ans: **(d)**

Q. 43. **For starting large DC motors, direct-on-line starters are not suitable. The reason is that:**

(a) Huge voltage drop may occur in the supply mains

(b) The motor may run away

(c) The starting current may be very much high

(d) The starting torque may be very low

Ans: **(a)**

Q. 44. **Which one is the method for controlling the speed of a DC motor?**

(a) Field control method

(b) Armature resistance control method

(c) Voltage control method

(d) All of the above

Ans: **(d)**

Q. 45. **While using a DC shunt motor, controlling of speed by armature resistance variation is best suited for:**

(a) Variable torque drive   (b) Constant torque drive

(c) Constant power drive   (d) Variable power drive

Ans: **(b)**

Q. 46. **In the Ward–Leonard method of speed control, the direction of rotation of the motor is reversed by:**

(a) Reversing the supply system terminals

(b) Reversing the motor and generator armature terminals

(c) Reversing the connections of generator field terminals

(d) None of the above

Ans: (c)

Q. 47. For which motor, the speed increases with load torque?

(a) Shunt motor only

(b) Cumulative compound motor

(c) Reluctance motor only

(d) Differentially compounded motor

Ans: (d)

Q. 48. What will be the effect on back emf, when the speed of a DC shunt motor is increased above its rated speed?

(a) It will decrease continuously

(b) It will increase continuously

(c) Remains unaffected

(d) It may increase/decrease according to the load

Ans: (c)

Q. 49. Field control in a DC shunt motor gives:

(a) Constant load speed drive

(b) Variable load speed drive

(c) Constant kW drive

(d) None of the above

Ans: (c)

Q. 50. What will be the effect on the speed of motor, if the applied voltage across its armature is increased by 5% only (keeping flux and load current fixed):

(a) Increases by about 5%   (b) Decreases by about 5%

(c) Increases by about 10%   (d) None of these

Ans: (a)

Q. 51. Whenever an additional resistance is connected in series with a DC series motor, then how its speed will be effected?

(a) Speed of the DC motor remains unaltered

(b) Speed of motor decreases

(c) Speed increases in a fast manner and reaches above the rated speed

(d) None of these

Ans: **(b)**

Q. 52. Two DC series motors are connected in series to drive the mechanical load. If these motors are now connected in parallel mode, then the new speed will be:

(a) Slightly more than double

(b) Zero

(c) Three times the previous speed

(d) Two times the rated speed

Ans: **(a)**

Q. 53. In the Ward–Leonard method of speed control of DC motor, the lower limit of speed is dependent upon:

(a) Field circuit resistance of motor

(b) Speed of the generator–prime mover

(c) Residual magnetism of the generator

(d) Both (b) and (c)

Ans: **(d)**

Q. 54. For controlling the speed of a DC shunt motor, above the base speed over a wide range, the motor must possess:

(a) Compensating winding

(b) Shunt winding with a rheostat

(c) Interpole winding and compensating winding both and must be started using a 4 point DC starter

(d) None of the above

Ans: **(c)**

Q. 55. Ward–Leonard method of speed control of DC motor is not used for:

(a) Wide range of speed

(b) Low speeds

(c) Constant speed operations

(d) None of these

Ans: **(c)**

Q. 56. In a non-interpolar machine, the speed can be increased by flux control method of DC shunt motor in the ratio:

(a) 4:3

(b) 2:1

(c) 1:2

(d) 3:4

Ans: (b)

Q. 57. Voltage control method for controlling the speed of DC shunt motor is applicable for:

(a) DC compound wound motor

(b) DC series motor

(c) Both (a) and (b)

(d) Neither (a) nor (b)

Ans: (c)

Q. 58. Armature or rheostatic control method for speed control of DC motors is preferred, when:

(a) Speed above no load speed are needed

(b) Speed below no load speed are needed

(c) Rated speed is required

(d) None of the above

Ans: (b)

Q. 59. Ward–Leonard method of speed control may be used for:

(a) Elevators, electric excavators

(b) Blooming and paper mills

(c) Cranes and hoists

(d) Both (a) and (b)

Ans: (d)

Q. 60. Which one of the following statements is not correct?

(a) A modification of Ward–Leonard system is called as Ward–Leonard Ilgner system

(b) The efficiency of Ward–Leonard method is low

(c) Ward–Leonard method provides a smooth accele-ration

(d) Ward–Leonard arrangement has only three motors and one generator coupled to each other

Ans: **(d)**

Q. 61. The speed of the DC motor increases in a fast manner at no load. It happens due to:

(a) Short circuit in supply mains

(b) Short circuit in field winding

(c) Open circuit in supply mains

(d) Open circuit in field winding

Ans: **(d)**

Q. 62. Which one of the following methods is suitable for speed control of series motor?

(a) Field diverter method

(b) Paralleling field coils method

(c) Tapped field control method

(d) All of the above

Ans: **(d)**

Q. 63. Which one of the following statements is wrong w.r.t. (rheostatic control method) speed control?

(a) Great amount of power is wasted in controller resistance

(b) It provides speed above the normal speed

(c) It requires costly overall arrangement

(d) In this case, efficiency is reduced

Ans: **(b)**

Q. 64. In series-parallel control of DC series motors, the torque is X times (in series mode) that created by motors, when in parallel. The value of X will be:

(a) 2   (b) 4

(c) 6   (d) 8

Ans: **(b)**

Q. 65. The speed of a DC shunt motor (more than its full load speed) may be obtained by:

(a) Reducing the current in the field winding

(b) Increasing the current in the armature winding

(c) Reducing the current in armature winding

(d) Any of the above

Ans: (a)

Q. 66. The speed of a DC shunt motor:

(a) Reduces to a small extent with the increment in load

(b) Increases to a large extent with the increment in load

(c) Remains same under all load conditions

(d) None of these

Ans: (a)

Q. 67. A DC series motor is drawing one ampere load current from the lines. Now the load is adjusted in such a way that the current drawn becomes half. In this situation, what will be the speed of the machine, if the saturation effects are neglected for the time being?

(a) It will be increased by 100%

(b) It will be reduced to 50% of previous value

(c) It will be reduced to 25% of previous value

(d) Remains unaltered

Ans: (a)

Q. 68. The speed regulation is defined as:

(a) $\dfrac{\left(N_0 - N_f\right)}{N_f}$
(b) $\dfrac{\left(N_f - N_0\right)}{N_f}$

(c) N_f / N_0
(d) None of the above

Ans: (a)

Q. 69. Which one of the following methods is termed as the most economical method of electric braking:

(a) Rheostatic braking

(b) Plugging

(c) Regenerative braking

(d) All of the above

Ans: **(c)**

Q. 70. Which type of electric braking is preferred for non-reversing DC drives?

(a) Dynamic braking

(b) Plugging

(c) Regenerative braking

(d) Dynamic braking with separate excitation

Ans: **(d)**

Q. 71. In DC motors, plugging is done by:

(a) Reversing the armature polarity

(b) Reversing the field winding polarity

(c) Reversing the polarities of both windings

(d) Any one of the above

Ans: **(a)**

Q. 72. Plugging is also known as:

(a) Counter current braking

(b) Counter emf braking

(c) Dynamic braking

(d) None of the above

Ans: **(a)**

Q. 73. In case of DC shunt motors, regenerative braking is applied, when the load:

(a) Is continuously decreasing

(b) Is continuously increasing

(c) Has an overhauling characteristic

(d) Is of resistive nature

Ans: **(c)**

Q. 74. Which type of braking provides highest braking torque?

(a) Plugging

(b) Regenerative braking

(c) Dynamic braking

(d) Both (b) and (c)

Ans: **(a)**

Q. 75. **In regenerative braking, the back emf is:**

(a) Greater than supply voltage

(b) Equal to supply voltage

(c) Lesser compared to supply voltage

(d) None of the above

Ans: **(a)**

Q. 76. **In a DC series motor, 20 N-m torque is developed, while drawing 3 A of load current. However, if the current reaches to 6 A, then what will be the amount of torque developed?**

(a) 80 N-m   (b) 20 N-m

(c) 30 N-m   (d) 60 N-m

Ans: **(a)**

Q. 77. **For DC series motors, the torque developed is proportional to:**

(a) I_a (b) I_a^2

(c) I_a^3 (d) $\sqrt{I_a}$

Ans: **(b)**

Q. 78. **When current is increased from 10 A to 12 A in a DC series motor, the percentage increase in the torque developed will be:**

(a) 40%   (b) 44%

(c) 36.6%   (d) 20.34%

Ans: **(b)**

Q. 79. **In a DC shunt motor, 54 N-m torque is developed at the armature current of 10 A. If the armature current is raised to 20 A, how much torque will be produced?**

(a) 108 N-m   (b) 105 N-m

(c) 206.57 N-m   (d) 70 N-m

Ans: **(a)**

Q. 80. A 220 V DC shunt motor runs at nearly 500 rpm, when the current flowing in the armature winding is 50 A. However, if the torque is doubled, what will be the corresponding new speed?

(a) 400 rpm   (b) 476 rpm

(c) 600 rpm   (d) 890 rpm

Ans: (b)

Q. 81. Testing of DC machines is necessary so as to obtain:

(a) Constant losses   (b) Efficiency

(c) Power losses   (d) All of the above

Ans: (d)

Q. 82. Which one of the following tests is termed as direct test?

(a) Swinburne's test   (b) Hopkinson's test

(c) Retardation test   (d) Brake test

Ans: (d)

Q. 83. In case of brake test, the efficiency of DC motor depends upon:

(a) Supply voltage

(b) Full load current taken by motor

(c) Spring balance weight

(d) All of the above

Ans: (d)

Q. 84. Swinburne's test is also known as:

(a) Losses method   (b) No load test

(c) Speed control test   (d) Both (a) and (b)

Ans: (d)

Q. 85. In Swinburne's method of obtaining the efficiency of a DC machine:

(a) The no load losses are measured and then copper losses are calculated

(b) Only copper losses are measured

(c) Only no load losses are measured

(d) The no load losses are calculated and then copper losses are measured

Ans: **(a)**

Q. 86. Which one of the followings is the most important advantage of Swinburne's test?

(a) It is applicable for those machines in which flux is practically constant

(b) It requires one running test

(c) It does not take care of iron losses

(d) It is very economical and convenient

Ans: **(d)**

Q. 87. The Swinburne's test is not found suitable for DC series machine, because:

(a) A series motor runs at dangerously high speed on no load

(d) A series motor suffers from commutation problem

(c) Both (a) and (b)

(d) None of the above

Ans: **(a)**

Q. 88. In case of Swinburne's test for the determination of efficiency of a DC shunt machine, the no load input power supplies:

(a) Shunt field copper losses

(b) Iron losses

(c) Armature losses and friction/windage losses

(d) All of the above

Ans: **(d)**

Q. 89. In Swinburne's test, the shunt machine is allowed to run as a:

(a) Generator at no load with rated speed and rated terminal voltage

(b) Generator at full load at any speed

(c) Generator at half load at rated speed

(d) None of the above

Ans: (a)

Q. 90. In case of Hopkinson's test, DC machines have:

(a) Equal speed

(b) Generator armature current, which is less than that in the motor

(c) Stray load loss in motor, which is more than that in generator

(d) All of the above

Ans: (d)

Q. 91. In Hopkinson's test:

(a) One machine runs as a motor and other acts as a generator

(b) The field current of the machine, operating as a generator is more than that of the machine operating as a motor

(c) Both the machines are operated at 7/10th of the rated load

(d) Both (a) and (b)

Ans: (d)

Q. 92. Which one of the followings is the main disadvantage of Hopkinson's test?

(a) It needs two identical DC shunt machines

(b) It requires a motor-generator set

(c) It is applicable to only shunt and compound machines

(d) All of the above

Ans: (a)

Q. 93. Which one of the following tests is not performed on shunt machine?

(a) Hopkinson's test   (b) Swinburne's test

(c) Retardation test   (d) Field test

Ans: (d)

Q. 94. Which one of the following tests is beneficial for obtaining efficiency of a DC traction motor?

(a) Retardation test   (b) Running down test

(c) Field test   (d) Brake test

Ans: **(c)**

Q. 95. In the field's test, for DC series machines, the series fields of two machines are connected in series so as to make.......losses of both machines equal:

(a) Friction and windage losses

(b) Copper losses

(c) Neither (a) nor (c)

(d) Iron losses

Ans: **(d)**

Q. 96. The main purpose of performing retardation test is to obtain:

(a) Stray losses

(b) Copper losses

(c) Iron losses

(d) Friction and windage losses

Ans: **(a)**

Q. 97. In running down test, the rate of loss of kinetic energy of armature is represented as:

(a) $\dfrac{d}{dt}\left(\dfrac{1}{2}Iw^2\right)$ (b) $\dfrac{d^2}{dt^2}\left(\dfrac{1}{2}Iw^2\right)$

(c) $\dfrac{d}{dt}\left(\dfrac{1}{2}Iw\right)$ (d) $\dfrac{d^2}{dt^2}\left(\dfrac{1}{2}Iw\right)$

Ans: **(a)**

Q. 98. Retardation test for DC machines may also be known as:

(a) Running down test   (b) Hopkinson's test

(c) Losses method   (d) Any one of the above

Ans: **(a)**

MISCELLANEOUS

Q. 1. A DC machine may provide maximum output at an efficiency of:

(a) 50%   (b) 20%

(c) 15%   (d) 27%

Ans: (a)

Q. 2. For a given torque, reduction in the diverter resistance of a DC series motor:

(a) Increases its speed, demanding more armature current

(b) Decreases its speed, demanding more armature current

(c) Decreases its speed, demanding less armature current

(d) None of the above

Ans: (a)

Q. 3. In the block diagram of separately excited DC motor, the armature induced emf appears as a:

(a) Negative feedback   (b) Input or output

(c) Positive feedback   (d) Disturbed output

Ans: (a)

Q. 4. In a DC separately excited motor, the developed torque (T) under constant terminal voltage is related with output power P as:

(a) $T^2 \propto P^{3/2}$ (b) $T \propto P^{3/2}$

(c) $T \propto P$ (d) $T^2 \propto \sqrt{P}$

Ans: (c)

Q. 5. What must be the ratio of output powers of two identical lossless series motors, running at N_1 and N_2 rpm and connected in series across a DC voltage?

(a) $N_2 : N_1$ (b) $N_1 : N_2$

(c) $N_2^3 : N_1^{3/2}$ (d) $N_2 : \sqrt{(N_1)}$

Ans: (b)

Q. 6. In DC machines, the space waveform of the air gap flux distribution affects:

(a) Neither the voltage nor the torque

(b) Only the voltage

(c) Only the torque

(d) Torque but not voltage

Ans: (a)

Q. 7. In DC machines, which one of the followings is the most powerful electromagnet, operating at full load condition?

(a) Interpole and compensating winding together

(b) Armature winding

(c) Field winding

(d) Only compensating winding

Ans: (c)

Q. 8. Whenever an AC supply is given to DC motor, the motor:

(a) Runs at a very high speed

(b) Runs at a very low speed

(c) Burns due to heat produced in its field by eddy currents

(d) None of these

Ans: (c)

Q. 9. In a DC series motor:

(a) Torque developed is poor

(b) Speed remains almost constant

(c) No starter is generally used for starting purpose

(d) Field winding consists of thicker wire having lesser number of turns

Ans: (d)

Q. 10. Which one of the following DC motors is used for shears and punches?

(a) Differential compound wound DC motor

(b) Cumulative compound wound DC motor

(c) Series motor

(d) All of the above

Ans: **(b)**

Q. 11. **Whenever the number of commutator segments of a DC machine is increased, then:**

(a) The shape of output wave becomes smooth

(b) Total output power is reduced

(c) Total output power remains nearly constant

(d) None of these

Ans: **(a)**

Q. 12. **What is the role of equalizer ring in a lap wound armature?**

(a) It reduces the noise even in the machine

(b) It helps to improve the speed of the machine

(c) It avoids unequal distribution of current at brushes

(d) It becomes helpful in starting of the DC motor

Ans: **(c)**

Q. 13. **Dummy coils are used in DC machines so as to:**

(a) Reduce harmonics problem

(b) Eliminate armature reaction

(c) Bring mechanical balance of armature

(d) None of these

Ans: **(c)**

Q. 14. **The nameplate of an electric motor indicates:**

(a) The output power available at the shaft

(b) The rated input power

(c) The gross value of power

(d) All of the above

Ans: **(a)**

Q. 15. **In DC machines, the armature core is made up of those materials, which have:**

(a) Small B-H loop area

(b) Large B-H loop area

(c) High value of resistivity

(d) Low value of resistivity

Ans: **(a)**

Q. 16. **Which type of DC motor is used for driving constant speed line shafting lathes?**

(a) DC cumulative compound motor

(b) DC shunt motor

(c) DC series motor

(d) None of the above

Ans: **(b)**

Q. 17. **Application of electrical machines depends upon:**

(a) Cost and ease of speed control

(b) Maintenance requirement

(c) Power factor

(d) Both (a) and (b)

Ans: **(d)**

Q. 18. **In case of DC machine, Aluminium is not used as a winding wire, because:**

(a) Aluminium conductors require a large winding space and produce jointing problems

(b) Aluminium has low value of conductivity

(c) Aluminium has very low value of resistivity

(d) None of these

Ans: **(a)**

Q. 19. **The direction of rotation of DC motor can be reversed by:**

(a) Reversing the terminals of armature winding

(b) Reversing the terminals of the field winding

(c) Reversing the connections of either the armature or the field winding with the supply

(d) Reducing the supply voltage

Ans: **(c)**

Q. 20. A DC series motor is never started without load because
(a) It will rotate at a dangerously high speed at no load
(b) It draws very small current at no load
(c) It draws very large current at no load
(d) None of these

Ans: (a)

Q. 21. A DC machine is running at a speed of 1200 rpm and the value of emf induced is 200 V. What will be the amount of electromagnetic torque developed if 15 A current is flowing in armature winding?
(a) 24 N-m　　　　　　(b) 23.9 N-m
(c) 30 N-m　　　　　　(d) 27 N-m

Ans: (b)

Q. 22. Which one of the followings is the reason of heavy sparking on the commutator?
(a) Motor overloaded
(b) Dirty commutator
(c) Brushes are not properly seated and also not in MNP position
(d) All of the above

Ans: (d)

Q. 23. A DC shunt motor produces 10 bhp at speed of 600 rpm. It draws 18 A at 500 V. What will be the values of efficiency and torque?
(a) 81.72%; 117 N-m　　(b) 85%; 117.72 N-m
(c) 89.23%; 117 N-m　　(d) 92%; 115 N-m

Ans: (a)

Q. 24. A DC machine has following data:
Ampere conductors = 25000, Length = 0.3 metre, Core diameter = 1.0 metre, Pole arc = 70%, Flux density in air gap = 0.6 T.

What must be the magnitude of torque developed by the armature?

(a) 1575 N-m   (b) 1675 N-m

(c) 1756 N-m   (d) 1726 N-m

Ans: **(a)**

Q. 25. **Which one of the following statements is wrong?**

(a) In DC series motors, starting torque is very high up to 500%

(b) In adjustable speed DC shunt motors, maximum momentary operating torque is usually limited to nearly 200% by commutation

(c) In constant speed DC shunt motors, the speed regulation is nearly 5 to 10%

(d) Differential compound wound DC motors are used for elevators and cranes

Ans: **(d)**

Q. 26. **When the two DC series motors are connected in series, the torque developed is T_s. After sometime, these are connected in parallel combination and the torque produced is T_p. However if mechanical power remains the same, then what will be the relation between these two torques?**

(a) $T_s = 4T_p$ (b) $T_p = 4T_s$

(c) $T_s = 3T_p$ (d) $T_p = 2.5T_s$

Ans: **(a)**

Q. 27. **In DC machines, mechanical losses occur due to:**

(a) Brush friction and bearing friction

(b) Resistance offered to armature rotation

(c) Brushes used

(d) Both (a) and (b)

Ans: **(d)**

Q. 28. **What is the maximum temperature limit for class B type in rotating electrical machines?**

(a) 1070 °C   (b) 130 °C

(c) 1920 °C   (d) 910 °C

Ans: **(b)**

Q. 29. Which one of the following statements regarding installation and preventive maintenance of DC machines is correct?

(a) Periodic greasing increases the life of bearings and provides trouble free service

(b) The commutator is regularly lubricated

(c) Dust may cause failure of DC machine

(d) Periodic operation of DC machine may cause failure of copper brushes

Ans: **(a)**

Q. 30. Which of the following factors improve commutation in a DC machine?

(a) Shifting of brushes in the opposite direction for motor

(b) High value of contact resistance for brushes

(c) Low value of contact resistance for brushes

(d) Both (a) and (b)

Ans: **(d)**

Q. 31. The shape of air gap flux-density wave at no load (in DC machines) will be:

(a) Sinusoidal

(b) Rectangular

(c) Sometimes rectangular and sometimes sinusoidal

(d) Flat-topped

Ans: **(d)**

Q. 32. For the speed control of DC shunt motor, a field regulator is used. For a constant load–torque, the speed will be minimum for a regulator resistance of:

(a) Zero ohms   (b) 100 ohms

(c) 2 Mega ohms   (d) 25 ohms

Ans: **(a)**

Q. 33. In case of a DC machine, hysteresis loss depends upon:

(a) Maximum value of flux density

(b) Frequency of magnetic reversals

(c) Volume and grade of iron

(d) All of the above

Ans: **(d)**

Q. 34. A cumulatively compounded long shunt machine, when operated as a motor, would operate as a:

(a) DC shunt motor

(b) Differentially compounded long shunt motor

(c) DC series motor

(d) None of these

Ans: **(b)**

FILL IN THE BLANKS

1. In DC machines, the field winding is on the ....

Ans: **Stator**

2. The field poles are made of a stack of .... plates, riveted together.

Ans: **Steel**

3. The concentrated field winding, when excited with direct current, creates alternate north and south poles, called .... construction.

Ans: **Hetropolar**

4. DC machines use .... coil winding.

Ans: **Closed**

5. In DC motors, emf induced is called as ....

Ans: **Counter or back emf**

6. In DC machines, yoke is made up from laminations so as to reduce the ....

Ans: **Eddy current losses**

7. The commutator is actually a group of .... segments, insulated from each other by .... sheets.

Ans: **Wedge-shaped copper; Thin mica**

8. Field winding is a .... winding on salient poles, bolted to the stator frame.

Ans: **Concentrated**

9. The function of yoke is to provide path for the ....

Ans: Pole flux

10. Pole core and pole shoes are made from thin laminations of .... to reduce eddy current losses.

Ans: Sheet steel

11. Brushes are made of .... for low voltage high current DC machines.

Ans: Copper-graphite

12. The two basic types of armature winding is known as lap and .... winding.

Ans: Wave

13. A DC machine may work as an .... energy converter.

Ans: Electromechanical

14. Whenever the field winding is excited by its own armature, the machine is known as ....

Ans: Self-excited DC machine

15. The shunt field may also be known as .... field.

Ans: Voltage operated

16. The effect of armature mmf, on the main field flux distribution in the air-gap is known as ....

Ans: Armature reaction

17. MNA is the axis, which is perpendicular to the flux passing through the ....

Ans: Armature

18. The bad effects of armature reaction include .... field flux.

Ans: Distortion and net reduction in

19. In DC machines, iron losses particularly in teeth are much greater on .... than on ....

Ans: Load; No load

20. In DC machines, the short air gap at the pole centre and longer air gap at the pole tips are kept only to limit the effect of cross-magnetizing armature mmf on the ....

Ans: Main pole flux

21. Interpoles are placed in between the ....

Ans: Main poles

22. The brush voltage drop in a DC machine will be nearly .... volts.

Ans: 2

23. In short shunt connection, the shunt field is connected across ....

Ans: Armature terminals

24. Some electrical machines, like cranes need .... starting torque.

Ans: High

25. In DC machines, copper losses in armature circuit are approximately ....% of the output.

Ans: 3

26. In DC motors, the applied voltage has to force current through .... against counter emf E_b.

Ans: Armature conductors

27. Emf is induced in DC motors and known as ....

Ans: Counter emf E_b

28. E_b acts like a ...., which makes the motor a self regulator.

Ans: Governer

29. The efficiency of motor is given by the ratio of power developed by the .... input.

Ans: Armature

30. Gross value of mechanical power developed by a motor is ...., when the back emf is equal to half the value of applied voltage.

Ans: Maximum

31. Torque means the .... moment of force about an axis.

Ans: Twisting

32. .... torque is that torque, which is available for useful work.

Ans: Shaft

33. The lost torque is the difference of ...., which is due to iron and friction losses of the motor.

Ans: **Armature and shaft torque**

34. The characteristic, which is drawn between speed and torque is known as ....

Ans: **Mechanical characteristic**

35. In those cases, where high value of starting torque is needed, such as for accelerating heavy masses, ... motors are preferred.

Ans: **DC series**

36. A .... motor must never be started without load.

Ans: **DC series**

37. Shunt motor is basically a .... motor.

Ans: **Constant speed**

38. For lathes, wood working machines and driving shafting, ... motor is preferred.

Ans: **DC shunt**

39. Speed torque characteristic may be obtained from T_a/I_a and .... characteristic.

Ans: **N/I_a**

40. Cumulative compound motors are those, in which series excitation .... the shunt excitation.

Ans: **Helps**

41. For rapid transit systems, ... motors are employed.

Ans: **DC series**

42. It is very difficult for a series motor to stall itself under heavy loading because it develops ....

Ans: **High overload torque**

43. In those machines, where interpoles are used, a ratio of (maximum to minimum speed) .... is used.

Ans: **6:1**

44. Rheostatic control method is preferred, when the speeds below .... speed are needed.

Ans: **The no load**

45. A modification of the Ward–Leonard system is called as ....

Ans: Ward–Leonard-Ilgner system

46. Ward–Leonard system provides low efficiency at .... loads.

Ans: Light

47. By using field diverters in DC motors, flux may be decreased, so the speed of the motor ....

Ans: Increases

48. For a DC series motor, if armature current is reduced due to armature diverter (for a given load torque), the value of flux must ....

Ans: Increase

49. The limitation of field control method is that the commutation becomes ...., because the effect of armature reaction is greater on a weaker field.

Ans: Unsatisfactory

50. At low speed, the motor may be connected in .... and for high speed, may be connected in ....

Ans: Series; Parallel

51. For the same rating, DC series motor has .... starting torque in comparison to a DC shunt motor.

Ans: Higher

52. Field copper losses are nearly ....% to ....% of full load copper losses.

Ans: 20; 30

53. .... braking uses generator action so as to stop the motor.

Ans: Rheostatic or Dynamic

54. Rheostatic braking provides .... torque compared to plugging.

Ans: Lesser

55. Plugging and rheostatic braking are also known as .... respectively.

Ans: Reverse current and Dynamic braking

56. In motors, the starting resistance R is connected in series with the .... and not with the complete motor.

Ans: Armature

57. The basic role of 'HOLD-ON' coil is to hold-on the arm in full .... position, when the .... is in normal operation

Ans: Running; Motor

58. A two point starter may be used for starting DC .... motor.

Ans: Series

59. In DC shunt motor, speed regulation may be between ....% to ....%.

Ans: 10; 15

60. Brake test is a .... test.

Ans: Direct

61. In the testing of DC machine, the load on the motor is adjusted till it carries its .... current (while using brake test).

Ans: Full load

62. The brake test is not preferred for .... sized DC machine.

Ans: Large

63. Whenever a DC machine is subjected to heavy overloads, the resultant field waveform will be extremely ....

Ans: Distorted

64. For a DC machine, magnetization curve is the relationship between air gap flux and .... mmf.

Ans: Field winding

65. Whenever the series motor is required so as to work under rigorously varying loads, then the diverter resistance must be of highly .... nature.

Ans: Inductive

66. The field flux control method is suitable for drives requiring .... torque at low speed.

Ans: Large

67. In direct method, for testing of DC machines, DC machine is subjected to .... load and the whole output power is wasted.

Ans: Rated

68. Swinburne's test cannot be performed on DC .... motor.

Ans: Series

69. In Swinburne's test, ... amount of power is needed for testing, even in large machines because only no load losses are to be supplied from supply mains.

Ans: Low

70. Regenerative method is also known as .... method.

Ans: Hopkinson's

71. In field's test, two identical .... machines are needed, which are coupled mechanically.

Ans: DC series

72. In field's test for DC series machine, the motor and generator fields are in series, so the .... losses in both machines are same.

Ans: Iron

73. Series motor has the tendency of attaining dangerous speed at .... condition.

Ans: No load

74. The function of interpoles is to assist .... by producing auxiliary or commutating flux.

Ans: Commutation

75. Running down test is also known as .... test and used for obtaining .... losses.

Ans: Retardation; Stray

76. Whenever the armature slows down with zero excitation, the energy of armature is used to overcome .... losses only.

Ans: Mechanical

77. .... brushes are employed in electric motors so as to give a path for the flow of current.

Ans: **Carbon**

78. Magnetic and mechanical losses are collectively called as .... losses.

Ans: **Stray**

79. Starter is not used generally in a very small sized DC motor, because it has .... resistance and .... moment of inertia.

Ans: **High; Low**

80. .... coil is also called as field failure coil.

Ans: **No voltage**

81. The commutator used in DC machine is assembled of .... segments.

Ans: **Copper**

82. The brushes applied on the commutator may be of radial, trailing and .... types.

Ans: **Reaction**

83. At standstill condition of rotor of DC motor, back emf is ....

Ans: **Zero**

84. An armature diverter is a variable resistance of .... capacity and .... ohmic value.

Ans: **High current; Low**

85. By using OLRC, ... protection is achieved.

Ans: **Overload**

86. .... relays (overload type) are preferred in DC motor starters.

Ans: **Magnetic**

87. No voltage release coil is connected across .... through shunt field winding (3 point starter).

Ans: **Supply**

88. The overload current depends upon the distance between the electromagnet and .... to be pulled.

Ans: **Strip**

89. The common types of starter (in a DC series motor) are the drum controller, face plate starter and .... starter.

Ans: **Liquid**

90. Negative terminal (4 point starter) is brought out so as to make the independent circuit of .... coil.

Ans: **No voltage release**

TRUE / FALSE

1. DC machine is a versatile device used for energy conversion.

Ans: **True**

2. In DC machines, armature winding is a concentrated winding, placed in the slots.

Ans: **False**

3. In DC systems, fabricated steel is preferred for small DC machines.

Ans: **False**

4. Leakage flux is the flux, which does not enter the armature (i.e. not useful).

Ans: **True**

5. Commutator used in DC machines is of smooth rectangular shape.

Ans: **False**

6. The constructional features of a DC motor and DC generator are different.

Ans: **False**

7. The armature core is separated from the field poles by a small air gap, which allows the armature to rotate freely.

Ans: **True**

8. In a DC motor, the function of commutator is to create bidirectional torque.

Ans: **False**

9. The spring (in the brush holder) puts required pressure on the carbon brushes, so that proper contact is maintained between the brushes and the commutator surface.

Ans: **True**

10. Conversion of AC induced in the armature circuit into DC at the output is known as commutation.

Ans: **True**

11. The armature winding is that part of DC machine, where emf is induced and force is developed to turn the rotor.

Ans: **True**

12. In DC machines, the distortion of main field flux affects the limit of successful commutation.

Ans: **True**

13. In DC motors, the resultant flux is strengthened at the trailing pole tips.

Ans: **False**

14. Sparking is produced at the brushes due to compensating winding, during armature reaction.

Ans: **False**

15. In DC motors, interpoles have a polarity opposite to that of the following main pole in the direction of rotation of armature.

Ans: **True**

16. The shunt motor is often considered as a variable speed motor.

Ans: **False**

17. Series motor is a constant flux machine.

Ans: **False**

18. A DC series motor must never run unloaded.

Ans: **True**

19. A DC series motor developes high value of torque at low speed and low torque at high speed.

Ans: **True**

20. DC shunt motor may be proved useful for traction purposes.

Ans: **False**

21. In a DC motor, emf generated may also be known as counter emf.

Ans: **True**

22. Heavy inrush of starting current, taken by the motor may result in sparking at the commutator.

Ans: **True**

23. For small sized DC motors, there is no need to connect starting resistance in the armature circuit.

Ans: **True**

24. Whenever wide range of speed by shunt field control is needed, then the 4 point starter is used.

Ans: **True**

25. In DC motor, holding magnet or coil may also be called as no volt release or low volt release.

Ans: **True**

26. In starter circuit, overload release is provided in parallel with the armature circuit.

Ans: **False**

27. In a three point DC starter, the field circuit and the hold coil are in parallel.

Ans: **False**

28. Whenever a little or no speed control is needed in a DC motor; only 4 point starter may be used.

Ans: **False**

29. In industries, push-button type automatic starters are used.

Ans: **True**

30. DC motors are most suitable for wide range speed control.

Ans: **True**

31. The principal advantage of speed control by varying the armature circuit resistance is that it gives higher efficiency and lower operational costs.

Ans: False

32. In shunted armature method of speed control; external resistances are inserted both in series and in parallel with the armature.

Ans: True

33. If the series motor is to work under rigorously varying loads, then the diverter resistance must be highly inductive.

Ans: True

34. In Ward–Leonard method of speed control, the efficiency at low speed is higher than that obtained by other methods of speed control.

Ans: True

35. In DC machines, the no load rotational losses are basically the iron losses at working flux.

Ans: False

36. Stray load losses in DC machines are generated by the distortion of the air gap flux due to the armature-reaction.

Ans: True

37. The sum of no load rotational loss and stray load loss is called as total rotational loss.

Ans: True

38. In case of a DC series motor, the brake must be sufficiently tight before the motor is switched on to the supply.

Ans: True

39. In direct testing method of DC machine, the friction torque at a particular setting of handwheels H_1 and H_2, remains constant.

Ans: False

40. Swinburne's test is an indirect test of testing DC machines.

Ans: True

41. DC series machine cannot be tested by Swinburne's method.

Ans: True

42. In Swinburne's test, the machine is run as a generator at rated voltage and speed.

Ans: False

43. In Swinburne's test, no account is taken of the change in iron loss from no load to full load.

Ans: True

44. Hopkinson's test may also be called as back-to-back test.

Ans: True

45. In the original Hopkinson's test, two similar DC series machines are mechanically coupled.

Ans: False

46. During regenerative test of DC machines, the supply voltage should be kept at the rated value and test should be carried out at rated speed.

Ans: True

47. In Hopkinson's test, large machines can be tested at rated load by drawing appreciable power from supply.

Ans: False

48. The main disadvantage of Hopkinson's test is that two identical DC shunt machines are required for testing.

Ans: True

49. Hopkinson's test is suitable only for manufacturers of large DC machines.

Ans: False

50. In field's test for DC machines, two identical DC shunt machines are needed.

Ans: False

51. Field test is also a regenerative test.

Ans: False

52. In field's test, accuracy depends on the accuracy with which the motor input and generator output are measured.

Ans: True

53. The field efficiency test of DC series motors overcomes the problem of obtaining readings at relatively light loads by connecting the series field of the motor in series with the generator armature.

Ans: False

54. A DC shunt motor may be employed in lathes.

Ans: True

55. In a DC shunt motor, whenever field winding is disconnected from the armature and connected to an external voltage source, it is called a separately excited motor.

Ans: True

56. Series motors may be employed to drive fan-load.

Ans: True

57. Series motors can withstand severe starting duties and can furnish high starting torques.

Ans: True

58. Compound motors with more steeper characteristics are employed, where load fluctuates between wide limits intermittently.

Ans: True

59. Permanent magnet DC motors may be used for blowers in heaters.

Ans: True

60. Mechanical braking for DC motors, requires frequent maintenance and replacement of brake shoes.

Ans: True

61. Reverse current braking may also be known as dynamic braking.

Ans: False

62. In reverse current braking, a braking resistor is connected across the field.

Ans: False

63. Rheostatic braking is also called as dynamic braking.

Ans: True

64. Plugging is a highly inefficient method of braking.

Ans: True

65. Dynamic braking is commonly preferred in rolling mills, machine tools and in controlling elevators.

Ans: False

66. For counter current braking, motor reversal may be done by reversing the polarity of the applied voltage either to the armature or to the field winding.

Ans: True

67. In general, dynamic braking is preferred for non-reversing drives as this simplifies the drive circuit.

Ans: True

68. In a DC series motor, dynamic braking may be obtained in two ways viz. with self-excitation and with separate excitation.

Ans: True

69. In DC series motors, regenerative braking is not possible because with the reversal of direction of armature current, the direction of field excitation also reverses.

Ans: True

70. Braking torque can be applied by friction brakes only.

Ans: False

71. Rheostatic braking of a DC series motor can also be affected by reconnecting it as a separately excited machine.

Ans: True

72. For reversing drives, plugging is useful for braking.

Ans: True

VIVA VOCE QUESTIONS

Q. 1. What do you mean by a DC motor?

Ans: DC motor is a machine, which converts electrical energy into mechanical energy.

Q. 2. On which principle, DC motor is based?

Ans: The principle of DC motor is that when a current carrying conductor is placed in a magnetic field, it experiences a mechanical force, whose direction is given by Fleming's left hand rule and the magnitude is given by $F = BIL$ Newton.

Q. 3. Which material is normally used to design the frame of a DC motor?

Ans: Cast iron

Q. 4. A DC motor having two poles only is sometimes used. By which name, it is generally known?

Ans: A bipolar machine

Q. 5. Which material is normally used for commutator?

Ans: Hard drawn copper

Q. 6. In the above question, which material is preferred for modern DC machines (now-a-days)?

Ans: Silvered copper

Q. 7. What do we mean by back emf in case of DC motor?

Ans: Whenever DC supply is given to a DC motor, motor armature or conductors rotate and cut the flux. Now due to the lack of electromagnetic induction, an emf is induced in them, opposite to the applied voltage according to Fleming's right hand rule.

Q. 8. What do you mean by voltage equation of the motor?

Ans. When a DC supply source is used across the armature then it has two functions
(a) It overcomes the counter emf E_b and
(b) Also supplies the armature ohmic drop ($I_a R_a$)
Thus, it can be said that
Applied voltage = Back emf + ohmic drop, i.e.
$V = E_b + I_a R_a$
It is called as the voltage equation of the motor.

Q. 9. What is the condition for obtaining maximum power in DC motor?

Ans: $E_b = V/2$

Q. 10. What do you mean by the term "torque" in a DC motor?

Ans: Torque signifies the twisting moment of a force about an axis. Also the direction of torque may be determined by Fleming's left hand rule.

Q. 11. On which factors, back emf depends?

Ans: It depends upon flux, number of armature, conductors, speed, number of poles and parallel paths in the system, hence given as

$E_b = \phi ZN/60 \times P/A$

Q. 12. What do you mean by armature and shaft torque in case of a DC motor?

Ans: The torque developed by armature is known as armature torque (T_a) but whole of this torque is not actually utilized for doing any useful work. Due to the iron and friction losses in a DC motor, the net torque available at the shaft is less compared to armature torque.

Thus, the torque, which is actually available for doing useful work is called shaft torque (T_{sh}). It is so called because it is available on the shaft.

Q. 13. What is the basic formula for the speed of a DC motor?

Ans: It is given as

$N = K E_b/\phi$,

where E_b is the counter emf and ϕ is the flux.

Q. 14. How can the direction of rotation of a DC shunt motor be reversed?

Ans: By reversing the current flow through either the armature winding or the field winding.

Q. 15. How can the direction of rotation of a DC series motor be changed?

Ans: It can be done by reversing the supply terminals to either armature or series field.

Q. 16. What is polar axis in a DC motor?

Ans: Polar axis is that axis, along which the main flux is set up.

Q. 17. What do we mean by brush axis?

Ans: The axis perpendicular to the polar axis is called the brush axis, along which the brushes are placed.

Q. 18. What is the main difference between cumulative compound and differential compound wound DC motors?

Ans: In cumulative compound wound DC motor, the series field assists the shunt field whereas in differential compound wound motor, series field opposes the shunt field.

Q. 19. What is the difference in the directions of electromagnetic torque developed in DC motor and DC generator?

Ans: In a motor, electromagnetic torque acts in the direction of rotation and in generator, it acts in opposite direction to the direction of rotation.

Q. 20. Why the DC motor is called as a "self regulating machine"?

Ans: Actually the presence of back emf makes the DC motor a self regulating machine, i.e. it makes the DC motor to draw as much armature current as is just sufficient to develop the required load torque.

Q. 21. What do you mean by the power equation of a DC motor?

Ans: In case of DC motor

$V = E_b + I_a R_a$

Now multiplying both sides by I_a, we have

$VI_a = E_b I_a + I_a^2 R_a$

where VI_a = power supplied to motor armature

$I_a^2 R_a$ = Power lost in armature

$E_b I_a$ = Power developed by motor armature

Q. 22. What do you mean by speed regulation of DC motor?

Ans: It is defined as the change in speed, when the load on the motor is reduced from full load to zero and is expressed in percentage of rated full load speed.

$$\% \text{ speed regulation} = \frac{\left(N_a - N_f\right)}{N_f} \times 100,$$

where N_a = No load speed
N_f = Full load speed.

Q. 23. What do you mean by BHP?

Ans: In electric motors, the mechanical power available at the shaft in horse power is called as brake horse power and written as BHP.

Q. 24. For which services, DC series motors are not suitable?

Ans: It is not suitable, where
(a) The load may be entirely removed.
(b) Driving is done by means of belt (because mishap to the belt would cause the motor run on no load)

Q. 25. Why the DC series motors are used for cranes and hoists?

Ans: Actually DC series motors create starting torque comparatively greater than that developed by a shunt motor for a given current. Thus series motors are best suited, where huge starting torque is needed such as for cranes and hoists, etc.

Q. 26. Which type of DC motor is suitable for

(a) Vaccum cleaner and
(b) Paper making mechanism

Ans: (a) DC series motor
(b) DC cumulative compound wound DC motor

Q. 27. Which type of DC motor is suitable for lathe machine?

Ans: DC shunt motor

Q. 28. Which type of insulation is used between the commutator segments of a DC machine?

Ans: Mica

Q. 29. What will happen, if back emf is not present in a DC motor?

Ans: If it happens, then its burning will take place.

Q. 30. In which portion of a DC motor, iron losses take place?

Ans: In the yoke

Q. 31. In a DC motor, on which factors mechanical output depends?

Ans: It depends both on back emf and armature current because we know that
$$P_{developed} = E_b I_a$$

Q. 32. For which system, lap winding is used in DC machines?

Ans: Lap winding is preferred for low voltage and higher current system.

Q. 33. What will happen, if the shunt field winding of a DC shunt motor suddenly breaks?

Ans: If it happens, then the speed of the motor will become dangerously high.

Q. 34. Which wire is used with interpole winding?

Ans: Thick copper wire

Q. 35. What must be the ratio of voltage applied to back emf, if maximum torque is in the a DC motor?

Ans: 2

Q. 36. What causes sparking at the brushes in a DC motor ?

Ans: Actually it happens due to the self-induction of the coil undergoing commutation.

Q. 37. Mention those reasons, due to which bearings become hot?

Ans: The reasons are as follows:
(a) Bearings may be not in line
(b) Lack of oil (lubrication)
(c) Belt may be too tight, etc.

Q. 38. In which applications, reverse current braking may be used?

Ans: It may be used in rolling mills, printing presses, machine tools, etc.

Q. 39. Which braking method is most economical?

Ans: Regenerative braking.

Q. 40. What is done actually in case of rheostatic braking?

Ans: In this braking method, the motor is disconnected from the supply and operated as a generator. Thus the kinetic energy of the DC machine is dissipated in resistance, connected in series with armature. Hence the complete phenomenon of electric braking takes place.

Q. 41. For which purposes, braking (while lowering loads) may be used?

Ans: It may be used for:
(a) Feeding power back to supply systems
(b) Controlling the speed at which the load comes down and limiting it to a safe value.

Q. 42. Brake test for testing DC machines is normally not used for large motors, why?

Ans: It happens because in large motors, it is difficult to dissipate the large amount of heat generated at the brake.

Q. 43. What are the various methods of determining the efficiency of DC machine?

Ans: These methods include:
(a) Direct method

(b) Indirect method

(c) Regenerative method

Q. 44. By employing indirect test for testing of DC machines, efficiency of which machines may be obtained?

Ans: By this method, shunt and compound motor's efficiency may be obtained.

Q. 45. What do you mean by brake test?

Ans: In brake test, by applying brakes (using suspended weights), net output of DC motor and efficiency may be obtained.

Q. 46. What is the main drawback of brake test in a DC machine?

Ans: The main drawback is that the output of motor cannot be measured accurately.

Q. 47. Why the Swinburne's test is known as indirect test?

Ans: In this method, first losses of machine are determined and hence efficiency at any load can be obtained. That's why it is called an indirect test.

Q. 48. What are the main advantages of Swinburne's test ?

Ans: The advantages are as follows:
(a) It is a convenient and economical method, because the power required so as to test a large machine is very small.
(b) Since constant losses are known, so the efficiency at any predetermined load can be obtained.

Q. 49. In Swinburne's test , constant losses may be taken equal to input at no load condition, why?

Ans: At no load condition, armature–copper losses are very small compared to other losses. That's why constant losses are taken nearly equal to input.

Q. 50. What is the most important disadvantage of Hopkinson's test?

Ans. The main disadvantage of this test is the necessity of two practically identical machines.

Q. 51. What are the advantages of Hopkinson's test?

Ans: The advantages include:

(a) The efficiency is being determined under load conditions so that the stray load loss is being taken into account.

(b) In this test, the power needed is small as compared to the full load power of two machines.

Q. 52. What is the purpose of retardation test?

Ans: This test is also called as running down test. The purpose is to find stray power loss of shunt wound DC machine.

Q. 53. What is the procedure for performing retardation test on a DC machine?

Ans: In this method of testing, the machine (under test) is accelerated slightly above its normal speed and the supply to the armature is cut off. Thus, the armature slows down and the kinetic energy of this armature is used to meet the rotational losses, i.e. stray power losses.

Q. 54. Why the large series machines cannot be tested by Swinburne's test?

Ans: Series motors can not be run on no load due to the problem of attaining dangerous high speeds.

Q. 55. For which type of DC motor, field's test is used?

Ans: DC series motor.

Q. 56. Why the field test is not called a regenerative test?

Ans: Since the output of generator is not fed back to motor, but is dissipated in a resistor, so field test is not a regenerative test.

Q. 57. What do you mean by stray power losses?

Ans: These are actually rotational losses. In other sense, it may be treated as the sum of magnetic and mechanical losses.

Q. 58. What is the main role of performing temperature run test on DC machines?

Ans: The main role is to find out the actual maximum temperature attained, while the machine is operating under certain load conditions.

Q. 59. What are the various methods for measurement of temperature rise?

Ans: These methods include:
(a) Resistance method
(b) Embedded temperature detector method
(c) Thermometer method

Q. 60. What is the limiting temperature for Class E insulation?

Ans: 120° C

Q. 61. Which type of faults may occur in field winding?

Ans: The faults may be open circuit, an earth fault or shorting of a coil either completely or some of its turns.

Q. 62. Which types of trouble may arise in case of DC motors?

Ans: It includes the following:
(a) Overheating, noise (excessive) and vibration
(b) Failure to start
(c) Sparking at brushes, etc.

Q. 63. At which point of a conductor, embedded in slot, does the maximum temperature occur?

Ans: It occurs at the centre of conductor.

Q. 64. On which factors do hysteresis and eddy current losses depend?

Ans: Actually iron loss (core loss) has two components namely hysteresis and eddy current losses. Hysteresis loss depends upon the maximum value of flux density attained in a magnetic cycle, frequency of magnetic cycles, volume of magnetic material and type of magnetic material. Likewise eddy current loss depends upon the thickness of magnetic material, laminations in addition to the factors mentioned above.

Q. 65. If P_1 and P_2 are full load copper loss and stray (including iron loss) loss respectively, then what must be the value of P_1/P_2? Consider that maximum efficiency occurs at 80% of full load.

Ans: 1.5625

Q. 66. What are the applications of separately excited DC motor?

Ans: It may be used for paper machines, diesel electric propulsion of ships, in steel rolling mills, etc.

Q. 67. Is shunt motor capable of starting heavy loads?

Ans: Yes, a shunt motor is capable of starting heavy loads but it would require more current input over normal values than the series and compound motors.

Q. 68. Why series motor is not suitable for services requiring a substantially constant speed?

Ans: It is so because, the speed of series motor falls rapidly (as according to its N/I_a characteristics) with the increase in load.

3

Electromechanical Energy Conversion

Q. 1. A circular iron core has an air gap cut in it and is excited by passing direct current through a coil wound on it. The magnetic energy stored in the air gap and the iron core is:

(a) In direct ratio of their reluctances

(b) In direct ratio of their path-lengths

(c) Equally divided between them

(d) None of these

Ans: (a)

Q. 2. A linear magnetic circuit has flux linkages of 1.2 Wb-turns, when a current of 10 A flows through the coil. What will be the energy stored in the magnetic field of this coil?

(a) 12

(b) 6

(c) 8

(d) 12.23

Ans: (b)

Q. 3. In an electromechanical energy conversion device, the torque developed depends upon:

(a) Stator and rotor field strength and torque angle

(b) Rotor field strength only

(c) Stator field strength only

(d) None of these

Ans: (a)

Q. 4. In a rotating electrical machinery, electromagnetic torque is established, when:

(a) Air gap is non-uniform

(b) Back up torque is less compared to pull out torque

(c) Both stator and rotor windings carry current

(d) None of these

Ans: **(c)**

Q. 5. The hysteresis loop area of a given magnetic material is 50 cm² with the two axes scaled as 1 cm = 20 AT and 1 cm = 50 mWb.

In case of 50 Hz frequency, the total hysteresis loss would be:

(a) 2.5 kW　　　　(b) 25 kW

(c) 35 kW　　　　(d) 3.5 kW

Ans: **(a)**

Q. 6. In an electric machine, slot leakage flux will be directly proportional to (where Z = conductors/slot, W = slot width):

(a) ZW　　　　(b) Z^2/W

(c) Z/W　　　　(d) Z/W^2

Ans: **(c)**

Q. 7. In an electromagnetic relay, force acts in a direction to:

(a) Minimize reluctance and maximize coil inductance

(b) Minimize reluctance only

(c) Minimize coil inductance only

(d) None of the above

Ans: **(a)**

Q. 8. Electromechanical energy conversion takes place via the medium of a:

(a) Magnetic field only

(b) Electric field only

(c) Magnetic field or electric field

(d) None of the above

Ans: **(c)**

Q. 9. Electromechanical energy converters are:

(a) Gross motion devices only

(b) Incremental motion devices only

(c) Either gross motion devices or incremental motion devices

(d) None of the above

Ans: **(c)**

Q. 10. Incremental motion devices include:

(a) Loudspeakers   (b) Microphones

(c) Electromagnetic relays (d) All of the above

Ans: **(d)**

Q. 11. The counter-clockwise torque produced due to inter-action of stator and rotor magnetic field is known as:

(a) The interaction or electromagnetic torque

(b) Reluctance torque

(c) Static torque

(d) None of the above

Ans: **(a)**

Q. 12. The current flowing through the coil sets up its own magnetic field, which interacts with main magnetic field, thus develops a torque called as:

(a) Electromagnetic torque

(b) Reluctance torque

(c) Dynamic torque

(d) All of the above

Ans: **(a)**

Q. 13. The torque developed in any motor is uniform in the direction:

(a) Perpendicular to magnetic axis

(b) Parallel line to magnetic axis

(c) Circumference of rotor

(d) Magnetic axis

Ans: **(c)**

Q. 14. Lenz's left hand rule of electromagnetism is applied for:

(a) DC motor only   (b) DC generator only

(c) Both   (d) None of the above

Ans: **(a)**

Q. 15. For generator action, total mechanical energy input will be equal to:

(a) Electrical energy output

(b) Total energy stored

(c) Total energy dissipated

(d) Summation of all of the above

Ans: **(d)**

Q. 16. With an electromechanical conversion device, the energy input in electrical form will be used as:

(a) Energy to electrical losses

(b) Energy to field storage in the electrical system

(c) Mechanical energy output

(d) Sum of all of the above

Ans: **(d)**

Q. 17. For a linear system, which statement is wrong?

(a) $Wf = \int_0^\psi i\, d\psi$ (b) $Wf = \int_0^\psi \dfrac{\psi}{L}\, d\psi$

(c) $Wf = \dfrac{1}{2}\int_0^\psi \dfrac{\psi^2}{L}\, d\psi$ (d) $Wf = \dfrac{1}{2L}\psi^2$

Ans: **(c)**

Q. 18. A solenoid is 0.3 m long and has a diameter of 0.03 m. It is wound with 1000 turns of wire and carrying a current of 10 A. The energy stored in the magnetic field will be?

(a) 0.025 J   (b) 0.015 J

(c) 0.35 J   (d) 0.25 J

Ans: **(b)**

Q. 19. Dynamo is a machine, which converts:

(a) Mechanical energy into electrical energy

(b) Electrical energy into mechanical energy

(c) Either (a) or (b)

(d) None of the above

Ans: **(c)**

Q. 20. Which one of the following statements is wrong?

(a) $W_{input} = W_{output} + W_{heat}$

(b) $\int dW_{input} = \int dW_{output} + \int dW_{heat}$

(c) $\dfrac{d\psi}{dt} = Bl\,\dfrac{dx}{dt} + Blv$

(d) None of the above

Ans: **(d)**

Q. 21. Electromechanical energy conversion devices include which of the following categories?

(a) Transducers

(b) Force producing devices

(c) Continuous energy conversion devices

(d) All of the above

Ans: **(d)**

Q. 22. Faraday's and Lenz's law are associated with which one of the following equations?

(a) $e = IR$ (b) $e = \dfrac{-d\psi}{dt}$

(c) $e = L\,dI/dt$ (d) All of the above

Ans: **(b)**

Q. 23. What will be the value of energy stored in the field, if the magnetic circuit has a linear B–H curve with $\lambda = 1.5$ WbT at $I = 10$ A?

(a) 30 J   (b) 7.5 J

(c) 25 J   (d) 27 J

Ans: **(b)**

Q. 24. In electromechanical energy conversion, magnetic effects may be provided by:

(a) Permanent magnets   (b) Electromagnets

(c) Reluctance variation   (d) All of the above

Ans: **(d)**

Q. 25. The hysteresis loop of a magnetic material has an area of 5 cm² with scales given as 1 cm = 2 A-T and 1 cm = 50 mWb. At 50 Hz, the total hysteresis loss will be:

(a) 25 W   (b) 2.5 W

(c) 35 W   (d) 35.02 W

Ans: **(a)**

MISCELLANEOUS

Q. 1. In a reluctance device (rotational), δ is the space angle between the axis of rotor and stator magnetic field. The developed torque will be proportional to:

(a) $\sin 2\delta$ (b) $\cos 2\delta$

(c) $2\,\delta^2$ (d) $\sin \delta$

Ans: **(a)**

Q. 2. In a physical system, electromagnetic force developed acts in such a direction that it tries to:

(a) Reduce the coenergy at constant mmf

(b) Increase the coenergy at constant mmf

(c) Decrease the coenergy at constant flux

(d) None of these

Ans: **(b)**

Q. 3. What will be the flux density at a point distant R, due to an infinitely long linear conductor carrying a current I?

(a) $B = \dfrac{\mu I}{2\pi R}$ (b) $B = \mu^2 I / 2R$

(c) $B = \mu I^2 / 2R$ (d) $B = IR/2\mu$

Ans: **(a)**

Q. 4. An electromechanical energy conversion device has a cylindrical stator but a salient pole rotor, which is not excited. If δ is the angle between stator field and rotor-long axis, then average torque developed will be proportional to:

(a) $A \sin (2\delta)$

(b) $A \sin (\delta)$

(c) $A \sin (2\delta) + B \sin (\delta)$

(d) $-A \cos (2\delta)$

Ans: (a)

Q. 5. In electromechanical energy conversion systems, magnetic field systems are employed, because:

(a) Stored energy density for tolerable electric field strength is low in electric field system

(b) Dielectric losses are large

(c) Stored energy density for tolerable electric field strength is high in electric field system

(d) None of the above

Ans: (a)

Q. 6. Singly and doubly excited magnetic systems are:

(a) Reluctance and synchronous motors respectively

(b) Induction motors

(c) Tachometers and transducers respectively

(d) DC shunt motors

Ans: (a)

Q. 7. In a single excited system, which statement is not considered as an assumption?

(a) Eddy current and hysteresis losses are neglected

(b) The magnetic field predominates and electric field effects are neglected

(c) Flux does not follow the magnetic path

(d) None of these

Ans: (c)

Q. 8. Doubly excited systems include:

(a) Tachometers

(b) Loudspeakers

(c) Seperately excited DC machines

(d) All of the above

Ans: **(d)**

Q. 9. A parallel plate capacitor has an electrode area of 100 mm², with a spacing of 0.1 mm between the electrodes. The dielectric between the plates is air with a permittivity of 8.85×10^{-12} F/m. The charge on the capacitor is 100 V. The stored energy in the capacitor is:

(a) 8.85 PJ

(b) 22.1 nJ

(c) 440 PJ

(d) 44.3 nJ

Ans: **(d)**

Q. 10. The energy stored in a magnetic field is given as:

(a) $1/2 \; AT^3$

(b) $3/2 \; Li^3$

(c) $1/2 \; \Phi^3 R$

(d) $1/2 \; Li^2$

Ans: **(d)**

FILL IN THE BLANKS

1. The electromechanical energy conversion device is a link between .... systems.

Ans: **Electrical and mechanical**

2. Electromechanical energy conversion takes place via the medium of ....

Ans: **Magnetic field or electric field**

3. Electromechanical energy converters may be either gross motion or .... devices.

Ans: **Incremental motion**

4. .... is a device, which converts mechanical energy into electrical energy.

Ans: **Generator**

5. .... is a device, which may convert electrical energy to mechanical energy.

Ans: **Motor**

6. Whenever a current carrying conductor is placed in a magnetic field, it experiences a ...., due to which torque may be produced.

Ans: **Force**

7. Electromechanical energy conversion devices may be classified into .... catagories.

Ans: **Three**

8. Energy conversion process is a .... process.

Ans: **Reversible**

9. Electromagnets, relays, actuators are the examples of .... producing devices.

Ans: **Force or torque**

10. Motors and generators are .... devices.

Ans: **Continuous energy conversion**

11. The analysis of energy conversion devices may be done by using .... energy concept.

Ans: **Field**

12. In energy conversion devices, the total amount of input energy is equal to the sum of energy stored, useful output energy and ....

Ans: **Energy dissipated**

13. Whenever any conductor rotates in a magnetic field, a voltage is .... in each conductor.

Ans: **Induced**

14. Electrical impulse measuring instruments, relays, etc. come under the category of .... devices.

Ans: **Incremental motion**

15. Coupling field is the link between .... systems.

Ans: **Electrical and mechanical**

16. The ferromagnetic rotor bears a .... urging it towards a region, where the magnetic field is stronger.

Ans: **Force**

17. The instantaneous value of energy stored in the magnetic field depends upon ....

Ans: Inductance and current

18. In a single excited system, it is assumed that the magnetic field .... and electric field effects are ....

Ans: Predominates; Neglected

19. For linear systems, generally half of the useful electrical energy input is stored in the magnetic field and the other half appears as .... output during the slow movement of rotor.

Ans: Mechanical energy

20. Tachometers, separately excited DC machines are the examples of .... system.

Ans: Doubly excited

TRUE/FALSE

1. Whenever a conductor is moved in a magnetic field, voltage is induced in the conductor.

Ans: True

2. An electromagnetic torque is produced on stator.

Ans: False

3. Reaction torque is produced, when the flux developed by the winding current interacts with stator only.

Ans: False

4. Energy conversion devices are needed at the ends of an electrical system.

Ans: True

5. Electromechanical energy conversion devices may be categorized into two types only.

Ans: False

6. Continuous energy conversion devices include motors and generators.

Ans: True

7. Coupling between the electrical and mechanical systems is through magnetic or electric field.

Ans: **True**

8. Reluctance torque may also be called as saliency torque.

Ans: **True**

9. In a doubly excited system, there are two independent sources of excitation.

Ans: **True**

10. The instantaneous values of energy stored in magnetic field depends on the inductance and current values at a particular instant.

Ans: **True**

11. The devices, which involve small motion, have only low energy signals from electrical to mechanical side.

Ans: **True**

12. Field energy concept is applicable to all the devices having rotational, linear or vibratory motion.

Ans: **True**

13. The energy storing capacity of magnetic field is nearly 25000 times greater than that of electric field.

Ans: **True**

14. The losses such as coupling field losses, friction and windage losses, etc. are irreversible type.

Ans: **True**

15. W_{fld} is a symbolic form, which represents the total energy absorbed by the coupling field.

Ans: **True**

16. Friction and windage losses come under the category of electrical system losses.

Ans: **False**

17. When a moving member can rotate or move with respect to a stationary member, the air gap must exist in between the stator and rotor.

Ans: **True**

18. When the output is in electrical form, then the coupling field must react with the mechanical system so as to absorb mechanical energy from it.

Ans: **True**

19. For a linear magnetic circuit, coenergy density depends upon H and B.

Ans: **True**

20. Doubly excited system is always preferred over singly excited system.

Ans: **False**

VIVA VOCE QUESTIONS

Q. 1. What is the basic purpose of (electromechanical, etc.) energy conversion?

Ans: Its purpose is to change the form of energy from one to another.

Q. 2. What do you mean by electromechanical energy conversion device?

Ans: It is a device, which converts electrical energy into mechanical energy or mechanical energy into electrical energy.

Q. 3. In how many categories, above mentioned devices may be categorized?

Ans: It may be divided into three categories:
(a) Devices involving small motion
(b) Force/torque producing devices
(c) Continuous energy conversion devices

Q. 4. What do you mean by continuous energy conversion devices?

Ans: These are the devices, which may convert one form of energy into another form, such as motor or generator. In case of generator, electrical energy is produced by mechanical energy and in motor, mechanical output is obtained by electrical energy.

Q. 5. Why the expressions for electromagnetic torque and generated voltage appear different for AC and DC machines?

Ans: Due to their different mechanical constructions.

Q. 6. What is the energy balance equation for generator?

Ans: It may be expressed as:

Total mechanical energy input = Electrical energy output + Total energy stored + Total energy dissipated

Q. 7. What is the basic meaning of energy dissipated regarding electromechanical energy conversion devices?

Ans: The total energy dissipated would be the summation of energy dissipated as I^2R losses in electrical system (+) energy dissipated in magnetic core as hysteresis and eddy current losses (+) energy dissipated in mechanical system as friction and windage losses.

Q. 8. What is the graphical representation of coenergy?

Ans: The area between the magnetization curve and the current or mmf axis is called as coenergy. It is denoted as W_f'.

Q. 9. For a linear magnetic system, what is the relation between field energy and coenergy?

Ans: Both these energies are represented as W_f and W_f' respectively. Thus it can be expressed as:

$$W_f = W_f' = \tfrac{1}{2}\, Li^2 = \frac{1}{2}\psi i = \frac{1}{2L}\psi^2$$

Q. 10. How many sources does the doubly excited system possess?

Ans: A doubly excited magnetic system has two independent sources of excitation.

Q. 11. What are the force/ torque producing devices?

Ans: These include relays, actuators, etc.

Q. 12. Whether energy conversion is a reversible process or irreversible process?

Ans: It is basically a reversible process.

Q. 13. Which quantities are analogous to charge and voltage in magnetic system?

Ans: These are analogous to flux linkage and current in the magnetic field.

Q. 14. Why the most converters prefer magnetic field as a coupling media between electrical and mechanical systems?

Ans: It is because the energy storing capacity of magnetic field is much greater than that of the electric field.

Q. 15. What is the condition for the development of useful torque in doubly excited systems?

Ans: Useful torque will be developed, when the angular velocity will be equal to the angular frequency of the supply current.

4

Single-Phase Induction Motor

Q. 1. Single-phase motors are:

(a) Easy to repair

(b) Cheaper in cost

(c) Simple in construction

(d) All of the above

Ans: (d)

Q. 2. A single-phase induction motor is:

(a) Inherently not self starting

(b) Self starting with the help of an auxiliary winding

(c) Self starting with the help of high capacitance only

(d) None of the above

Ans: (b)

Q. 3. A rotating magnetic field is produced by current in two windings displaced by 90 electrical degrees. It is the principle of:

(a) Phase splitting   (b) Phase advancing

(c) Both (a) and (b)   (d) None of the above

Ans: (a)

Q. 4. In a single-phase induction motor, the no load current is nearly x% of full load current. What is the value of x?

(a) 3.00%   (b) 2.00%

(c) 50%   (d) 40%

Ans: (d)

Q. 5. Single-phase induction motors may be employed:

(a) In vacuum cleaners

(b) In refrigerators

(c) In food processors

(d) All of the above

Ans: **(d)**

Q. 6. In single-phase induction motor, cage winding takes place on:

(a) Rotor

(b) Stator

(c) Neither rotor nor stator

(d) Both (a) and (b)

Ans: **(a)**

Q. 7. The performance of single-phase induction motor may be analysed from:

(a) Double revolving field theory

(b) Cross field theory

(c) Both (a) and (b)

(d) None of the above

Ans: **(c)**

Q. 8. In a double revolving field theory, what will be the slip of the backward motor, if the slip of the forward motor is S?

(a) $2 - S$ (b) $S - 2$

(c) $2S$ (d) $5S$

Ans: **(a)**

Q. 9. In a single-phase induction motor, the stator winding is splitted into two parts so as to:

(a) Develop starting torque

(b) Reduce speed

(c) Increase power factor

(d) Decrease power factor

Ans: **(a)**

Q. 10. As compared to 3-phase induction motors, single-phase induction motors have:

(a) Poor speed regulation   (b) Lower power factor

(c) Neither (a) nor (b)   (d) Both (a) and (b)

Ans: **(d)**

Q. 11. According to the process of producing starting torque, induction motors are further classified as:

(a) Shaded pole motors

(b) Repulsion start induction motors

(c) Split phase motors

(d) All of the above

Ans: **(d)**

Q. 12. The disadvantages of single-phase motors include:

(a) Low overload capacity

(b) Low power factor

(c) Low output

(d) All of the above

Ans: **(d)**

Q. 13. In single-phase induction motor, the winding having heaviest wire is called as:

(a) Auxiliary winding   (b) Compensating winding

(c) Main winding   (d) Tertiary winding

Ans: **(c)**

Q. 14. Which one of the following statements is incorrect?

(a) Single-phase induction motor may be employed in washing machines and refrigerators

(b) Single-phase induction motor is inherently not self starting

(c) In single-phase induction motor, a stationary pulsating magnetic field can be resolved into two rotating magnetic fields

(d) None of these

Ans: **(d)**

Q. 15. **When the rotor of single-phase induction motor is stationary:**

(a) Induced voltages will be equal and opposite

(b) Two torques will be equal

(c) Two torques will be zero

(d) Both (a) and (c)

(e) Both (a) and (b)

Ans: **(e)**

Q. 16. **The equivalent circuit, induced torque and torque-speed characteristic can be obtained by:**

(a) Cross field theory

(b) Pulsating field theory

(c) Two-reactance theory

(d) Double revolving field theory

(e) Either (a) or (d)

Ans: **(e)**

Q. 17. **During starting period of single-phase motor, it is converted as:**

(a) 3-phase motor   (b) Two-phase motor

(c) Repulsion motor   (d) Both (b) and (c)

Ans: **(b)**

Q. 18. **A 3-phase induction motor is running and supplying a constant load. Suddenly the fuse of one of the lines blown off. Now the machine is running as a single-phase induction motor. In this situation, line current value will increase to approximately:**

(a) 2 times   (b) $\sqrt{3}$ times

(c) 3 times   (d) $\sqrt{7}$ times

Ans: **(b)**

Q. 19. **Whenever relative comparison is made between poly-phase induction motor and single-phase induction motor with respect to torque-slip characteristic, it is found that when slip is zero:**

(a) Torque will be zero in case of polyphase induction motor but has a non-zero negative value on the single-phase induction motor

(b) Torque will be of non-zero negative value for both the machines

(c) Nothing can be said

(d) None of the above

Ans: **(a)**

Q. 20. In the double revolving field theory, the sinusoidally distributed mmf in the air gap is divided into...... components rotating in just opposite directions:

(a) Four   (b) Three

(c) Six   (d) None of the above

Ans: **(d)**

Q. 21. Which one of the following tests must be performed on single-phase induction motor so as to determine parameters?

(a) Blocked rotor test

(b) No load test

(c) Either (a) or (b)

(d) Both (a) and (b)

Ans: **(d)**

Q. 22. While performing blocked rotor test, what will be the value of slip?

(a) S = 1

(b) S = 0.2

(c) Varies between 0.2 to 1

(d) None of the above

Ans: **(a)**

Q. 23. The net induced torque in the single-phase motor will be equal to the (considering backward and forward torques):

(a) Difference between these torque values

(b) Summation of these torque values

(c) Multiplication of these torque values

(d) Nothing can be said

Ans: **(a)**

Q. 24. The rotational losses mean the summation of:

(a) Windage loss and core loss

(b) Copper loss and friction loss

(c) Windage loss, core loss and copper loss

(d) Friction loss, windage loss and core loss

Ans: **(d)**

Q. 25. Which one of the following statements is not true?

(a) In blocked rotor test, the rotor is at rest position

(b) The value of rotor slip with respect to forward rotating flux will be S (slip)

(c) Single-phase motors are employed for small power tools and blowers

(d) None of these

Ans: **(d)**

Q. 26. At no load test, the motor is allowed to run without load at:

(a) Rated voltage

(b) Rated voltage and rated frequency

(c) Twice the supply frequency

(d) Rated voltage and thrice the supply frequency

(e) Both (a) and (c)

Ans: **(b)**

Q. 27. Which one of the following motors in not a single-phase induction type motor?

(a) Reluctance start type

(b) Shaded pole type

(c) Compound wound motor

(d) Split phase type

Ans: **(c)**

Q. 28. Performance characteristics of single-phase induction motors are ...... compared to 3-phase induction motors:

(a) More satisfactory

(b) Less satisfactory

(c) Very much poor

(d) None of the above

Ans: (b)

Q. 29. In single-phase induction motor, average torque at standstill condition is:

(a) Zero

(b) 1.5 times full load torque

(c) 2.5 times full load torque

(d) 0.15 times full load torque

Ans: (a)

Q. 30. In split phase method, starting winding is displaced in space with the main winding by an angle of:

(a) 30°   (b) 90°

(c) 0°   (d) 120°

Ans: (b)

Q. 31. The auxiliary winding is connected:

(a) In parallel to the main winding across a single-phase supply

(b) In series with main winding

(c) In series with a protective relay always

(d) In parallel with a contactor type switch

Ans: (a)

Q. 32. Another name, frequently used for auxiliary winding is:

(a) Main winding

(b) Starting winding

(c) Compensating winding

(d) Lap winding

Ans: (b)

Q. 33. **Which one of the following motors is called single value capacitor motor?**

(a) Shaded pole motor

(b) Permanent split capacitor motor

(c) Capacitor start motor

(d) Split phase motor

Ans: **(b)**

Q. 34. **The split phase motor has a high starting current, which is nearly ….. times the full load value:**

(a) 5 to 10 times   (b) 7 to 8 times

(c) 2 to 3 times   (d) 2 to 7 times

Ans: **(b)**

Q. 35. **In split phase motor, the direction of rotation of resistance start induction motor may be reversed by reversing the line connections of:**

(a) Main winding

(b) Starting winding

(c) Either main winding or starting winding

(d) None of the above

Ans: **(c)**

Q. 36. **In split phase resistance start motor, the auxiliary winding is cut out of the circuit, when the motor has reached nearly……% of synchronous speed:**

(a) 40 to 70   (b) 20 to 40

(c) 50 to 90   (d) 70 to 80

Ans: **(d)**

Q. 37. **In permanent split single value capacitor motor, the capacitor is connected in:**

(a) Parallel with auxiliary winding

(b) Series with auxiliary winding

(c) Parallel with main winding

(d) Series with main winding

Ans: **(b)**

Q. 38. A two value capacitor motor has:

(a) Two capacitors, one is electrolyte type capacitor and other is oil type capacitor

(b) Same starting torque compared to a capacitor start motor

(c) Better running power factor compared to a capacitor start motor

(d) All of the above

Ans: **(d)**

Q. 39. In shaded pole method, the short circuited coils are also known as:

(a) Shading coils   (b) Shading rings

(c) Compensating coils   (d) Both (a) and (c)

(e) Both (a) and (b)

Ans: **(e)**

Q. 40. In a shaded pole method, the rotor of motor runs from the:

(a) Unshaded portion to shaded portion

(b) Shaded portion to unshaded portion

(c) Synchronous speed to 3600 rpm

(d) None of the above

Ans: **(a)**

MISCELLANEOUS

Q. 1. A single-phase induction motor is running at a fixed speed with main winding only. Which of the following statements regarding impedance is true?

(a) Impedances seen by forward and backward fields are equal

(b) Impedance seen by the forward field is much greater compared to that seen by backward field

(c) Both impedances will be nearly zero

(d) None of the above

Ans: **(b)**

Q. 2. Which motor is more suitable for computer printer drive?

(a) Stepper motor   (b) Shaded pole motor

(c) Hysteresis motor   (d) None of the above

Ans: (a)

Q. 3. A single-phase motor having single winding possesses:

(a) Zero starting torque

(b) Very high starting torque

(c) Very low starting torque

(d) None of the above

Ans: (a)

Q. 4. In a resistance split phase induction motor, when a centrifugal switch fails to close (when the motor is de-energized), then:

(a) There will be no starting torque when supply is connected again across the motor terminals

(b) Motor will develop dangerously high torque at the position of restart mode

(c) Motor will produce very low amount of torque at the position of restart mode

(d) Motor current will be very low at restarting situation

Ans: (a)

Q. 5. When a single-phase supply is connected across a single-phase motor, the magnetic field produced will be:

(a) Of rotating nature

(b) Pulsating in nature

(c) Rotating at twice synchronous speed

(d) Rotating at $\dfrac{Ns}{2}$

Ans: (b)

Q. 6. The direction of rotation of a shaded pole motor:

(a) Can not be reversed

(b) Can be reversed by connecting a capacitor in series

    (c) Can be reversed by connecting a resistor in series

    (d) None of the above

Ans: (a)

Q. 7. The capacitor used in a capacitor start induction motor has:

    (a) No polarity marking   (b) No voltage rating

    (c) Both (a) and (b)   (d) None of the above

Ans: (a)

Q. 8. In a capacitor-start induction motor, the capacitor is replaced by an inductor of equivalent reactance. Now the motor:

    (a) Will give humming sound

    (b) Will not start at all

    (c) Will run at very fast speed

    (d) Will run slowly

Ans: (b)

Q. 9. In a capacitor-start induction run motor, starting torque is related to angle α as:

    (a) $\sin 2\alpha$ (b) $\cos 2\alpha$

    (c) $\cos \alpha$ (d) $\sin \alpha$

Ans: (d)

Q. 10. If the auxiliary winding is left in the circuit, it will:

    (a) Draw excessive current (b) Run faster

    (c) Get overheated   (d) Both (a) and (c)

Ans: (d)

Q. 11. Which one of the followings is the disadvantage of shaded pole motor?

    (a) It has very low starting torque

    (b) It has little overload capacity and low efficiency

    (c) Both (a) and (b)

    (d) None of the above

Ans: (c)

Q. 12. Shaded pole motor may be used in:
    (a) Small fans   (b) Toys
    (c) Ventilators   (d) Phonograph turn tables
    (e) All of the above

Ans: (e)

Q. 13. In shaded pole single-phase motor, the required phase splitting is produced by:
    (a) Conduction
    (b) Induction
    (c) Both (a) and (b)
    (d) None of the above

Ans: (b)

Q. 14. The applications of split phase machine are:
    (a) In centrifugal pumps and seperators
    (b) In duplicating machines
    (c) In oil burners
    (d) All of the above

Ans: (d)

Q. 15. The equivalent circuit of a single-phase induction motor can be developed on the basis of:
    (a) Two reactance theory
    (b) Two revolving field theory
    (c) Salient pole theory
    (d) Non salient pole theory

Ans: (b)

Q. 16. Which of the following statements is not true regarding centrifugal switch?
    (a) It is mounted inside the motor
    (b) It consists of two parts, i.e. stator and rotor
    (c) It has the actuating mechanism, the centrifugally operated component
    (d) It remains close in all running positions

Ans: (d)

Q. 17. **In single-phase induction motors, the main winding occupies nearly:**

(a) 1/3rd of pole pitch

(b) 2/3rd of pole pitch

(c) 3/7th of pole pitch

(d) 1/2 pole pitch

Ans: **(b)**

Q. 18. **Split phase motors are convertible for use with:**

(a) 115 volts only

(b) 220 volts only

(c) Either 115 volts or 230 volts source of supply merely by changing the circuit connections

(d) None of the above

Ans: **(c)**

Q. 19. **Capacitor-start motors are prepared in rating ranging from:**

(a) 1/10 kW to 3/4 kW

(b) 1/20 kW to 3/40 kW

(c) 0.2 kW to 2 kW

(d) 1/10 kW to 2 kW

Ans: **(a)**

Q. 20. **Phase splitting can be done:**

(a) By causing the auxiliary winding to have low resistance

(b) By adding a capacitor in series with the auxiliary winding

(c) Both (a) and (b)

(d) Either (a) or (b)

Ans: **(d)**

Q. 21. **A capacitor start single-phase induction motor develompes greater starting torque compared to a split phase motor. It happens because the use of capacitors in the auxiliary winding enables:**

(a) The torque-slip characteristic in general to obtain a shape to give a large starting torque

(b) Provision of very large number of turns in the auxiliary winding

(c) Provision of large starting current

(d) None of the above

Ans: **(a)**

Q. 22. **The peripheral distance between two adjacent poles, is known as:**

(a) Pole pitch   (b) Coil pitch

(c) Coil span   (d) All of the above

Ans: **(a)**

Q. 23. **A multiturn coil possesses:**

(a) Five coil sides

(b) Two coil sides

(c) Three coil sides

(d) Six coil sides

Ans: **(b)**

Q. 24. **Which one of the following bearings is mostly used in fractional horse power motors?**

(a) Plain bearing

(b) Plain or sleeve bearing

(c) Ball and sleeve bearings

(d) Roller and sleeve bearings

Ans: **(b)**

Q. 25. **Which one of the following parameters will be needed so as to obtain the value of capacitor for a capacitor-start induction motor?**

(a) Maximum torque of motor

(b) Starting torque of motor

(c) Slip

(d) Speed of rotor

Ans: **(b)**

FILL IN THE BLANKS

1. A single-phase induction motor is similar to a 3-phase .... induction motor in physical appearance.

Ans: Squirrel cage

2. A single-phase induction motor has two parts namely stator and ....

Ans: Rotor

3. A single-phase induction motor has .... overload capacity.

Ans: Low

4. A non-uniform magnetic field produces a nonuniform .... which makes the operation of motor noisy.

Ans: Torque

5. Pulsating field means that the field builds up in one direction, reduces to .... and then builds up in the .... direction.

Ans: Zero; Opposite

6. The behaviour of a 1–φ motor may be explained by .... and .... theory.

Ans: Two field or double revolving field; Cross field

7. While dealing with double revolving field theory, an alternating sinusoidal flux can be represented by .... fluxes.

Ans: Two revolving

8. There is no .... torque in a single-phase induction motor.

Ans: Starting

9. During starting instant, 1–φ motor is temporarily converted into a .... motor.

Ans: 2 phase

10. In a split phase motor, the value of starting torque is proportional to ....

Ans: Sin α

11. Standard split phase motors have speed regulation, which is nearly .... as of the 3φ motors.

Ans: **Same**

12. In capacitor-start induction-run motors, a capacitor is connected in .... with the starting winding.

Ans: **Series**

13. In a split-phase motor, developed torque is proportional to the sine of the angle between starting current and ....

Ans: **Main current**

14. In single-phase motor, core loss can be represented by an equivalent ....

Ans: **Resistance**

15. In case of double revolving field theory, the .... direction is the direction in which the 1φ motor is started initially.

Ans: **Positive**

16. In the normal operation of 1φ motors, the auxiliary winding is disconnected from the supply, when the machine reaches a .... value.

Ans: **Certain**

17. In 1φ induction motor, the induced torque is equal to the ratio of .... and ....

Ans: **Air gap power; Synchronous angular velocity**

18. The parameters of the equivalent circuit of a 1φ induction motor can be obtained from the .... and ....

Ans: **No load; Blocked rotor tests**

19. During blocked rotor test, ... voltage is given to stator winding, so that .... current flows in main winding.

Ans: **Low; Rated**

20. In no-load test, the motor is allowed to run at .... voltage and frequency.

Ans: **Rated**

21. Permanent split capacitor motor is also known as .... motor.

Ans: **Single value capacitor**

22. Split phase induction motor consists of a …….. rotor and …….. having two windings.

Ans: Single cage; Stator

23. In resistance split phase motor, the main winding current lags behind the supply voltage (V) by nearly …….. (angle).

Ans: 90°

24. In split phase induction motor, a …….. switch can be used to disconnect the auxiliary winding for motors above 100 watts.

Ans: Centrifugally operated

25. In split phase motor, the pull out torque is nearly …….. times full load torque at nearly …….. % of synchronous speed.

Ans: 2.5; 75

26. Split phase motors are suitable for easily started loads where …….. is limited.

Ans: Frequency of starting

27. In capacitor start motor, single-phase current is allowed to split into …….. phases to be applied to the stator windings.

Ans: Two

28. Capacitor start motors are more …….. compared to split phase motors due to the additional cost of ………. .

Ans: Costly; Capacitor

29. In a two value capacitor motor, two capacitors are joined in …….. at starting.

Ans: Parallel

30. Permanent split capacitor motor has no …….. switch.

Ans: Starting

31. PSC motors have greater power factor due to ………. .

Ans: Permanently connected capacitor

32. In 1φ induction motors, 1φ …….. winding is employed.

Ans: Double layer

33. Whenever single-phase supply is connected across the 1φ stator winding, an …….. field will be produced.

Ans: Alternating

34. In split-phase resistance-start motor, auxiliary winding is placed at .... angle with the stator main winding.

Ans: 90°

35. Split phase induction motors, can be designed to develop starting torques in the range .... % of full load torque.

Ans: 150 to 200

36. The direction of rotation of split phase motor can be reversed by reversing the terminal connections of either .... or ....winding.

Ans: Main; Auxiliary

37. Whenever a large starting torque is needed to be developed, ... capacitors are used.

Ans: Two

38. When a single winding, single-phase motor runs in a particular direction, ... field is stronger compared to forward field.

Ans: Backward

39. In a 1φ induction motor, phase splitting can be accomplished by adding a .... in series with auxiliary winding.

Ans: Capacitor

40. The number of winding turns of auxiliary winding is kept low in a .... motor to keep the R/X ratio high.

Ans: Resistance split phase

41. In a capacitor-start motor, the starting torque lies in the range of .... to .... times its full load torque.

Ans: 3; 4

42. Shaded pole motors are preferred, where the starting torque requirement of the load is ....

Ans: Very low

43. Shaded pole is a pole on which a .... is fixed.

Ans: Shading ring

44. In a single voltage, three lead reversible type motor, a two-section .... winding is used.

Ans: Running

45. The motors, which start with …….. value of capacitance and run with a …….. value of capacitance are called as …….. capacitor-run motors.

Ans: High; Low; Two value

46. In shaded pole, 1φ motor, the required phase splitting is produced by …….. .

Ans: Induction

47. In shaded pole motors, efficiency varies from ……..% to …….. % (for higher ratings)

Ans: 5; 35

48. Shaded pole motors are built in sizes from 1/250 hp to …….. hp.

Ans: 1/6

TRUE / FALSE

1. Single-phase induction motor has no inherent starting torque.

Ans: True

2. Single-phase motors are basically large kilowatt rating motors.

Ans: False

3. Single-phase motors may only be classified on the basis of their construction.

Ans: False

4. The rotor of any 1φ motor is interchangeable with that of a polyphase induction motor.

Ans: True

5. If initial rotation is given to rotor in any particular direction, then speed emf is induced in it.

Ans: True

6. No starting arrangement is needed for starting single-phase induction motor.

Ans: False

7. The capacitor, which is connected in series with starting winding, improves overall power factor of the motor.

Ans: True

8. A permanent capacitor motor possesses greater efficiency compared to resistance-start induction motor.

Ans: True

9. At full load condition, permanent capacitor motor has low value of power factor.

Ans: False

10. When two windings are connected in parallel to a 1–φ supply source, the field (developed) will revolve.

Ans: False

11. Permanent capacitor motor may also be called as two value capacitor motor.

Ans: False

12. There exists nonuniform air gap between stator and rotor.

Ans: False

13. Single-phase induction motors are normally provided with concentric coils.

Ans: True

14. Single-phase motors have high efficiency and high value of power factor.

Ans: False

15. The behavior of single-phase induction motor may be explained by cross field theory only.

Ans: False

16. When the rotor of single-phase motor is allowed to rotate at N (speed), two slips will be S and $2 - S$.

Ans: True

17. In a single-phase motor, the increase in rotor resistance increases the effectiveness of the backward field.

Ans: True

18. Single-phase motors make more noise compared to 3-phase induction motors.

Ans: True

19. The speed regulation of 1φ motors tend to be better compared to polyphase motors.

Ans: **False**

20. Blocked rotor test is performed at nearly 5% of rated voltage so as to avoid heating.

Ans: **False**

21. No load test is performed at rated voltage for 1φ induction motor.

Ans: **True**

22. Centrifugal switch used in 1φ induction motor is mounted outside the motor.

Ans: **False**

23. In high resistance start motors, auxiliary winding is designed to remain energised for more than a few seconds.

Ans: **False**

24. In single-phase motors, the main winding occupies nearly 1/3rd of pole pitch.

Ans: **False**

25. In resistance-start 1φ induction motor, main winding current (I_m) lags behind the applied voltage V by 30°–45°, at starting.

Ans: **False**

26. Resistance start split phase motors give speed ranging from 287 to 700 rpm.

Ans: **True**

27. The capacitor used in the capacitor-start motor is of electrolytic type.

Ans: **True**

28. Increment in phase displacement (θ) between I_m and I_s decreases the starting torque of 1φ motor.

Ans: **False**

29. Fractional kW motors are normally single-phase type motors.

Ans: **True**

30. In capacitor-start 1φ induction motor, the capacitor size does not depend upon the rating of motor.

Ans: False

31. In permanent capacitor single-phase induction motor, no centrifugal switch is used.

Ans: True

32. In permanent capacitor 1φ induction motor, the same capacitor is used for starting and running.

Ans: True

33. In shaded pole motor, the rotor is a simple squirrel cage type with a 30° skew.

Ans: False

34. In shaded pole motor, the field flux shifts from shaded portion to unshaded portion.

Ans: False

35. By changing the applied voltage, the speed of a shaded pole motor may be changed.

Ans: True

36. The capacitor split phase motor provides high starting torque and low value of starting current.

Ans: True

37. In shaded pole machine, the shading ring can not be open-circuited (when the motor has started).

Ans: True

38. In ceiling fans, capacitance lying between 2 and 5 micro-farads remain permanently connected in series with the auxiliary winding.

Ans: True

39. The capacitor, in a two-value capacitor motor (for running purpose) is a dry-type AC electrolytic capacitor.

Ans: False

40. In a shaded pole squirrel cage induction motor, the flux in the shaded part always leads the flux in the unshaded segment.

Ans: False

VIVA VOCE QUESTIONS

Q.1. Why a single-phase motor is not self starting?

Ans: When 1φ supply is given to the stator winding, flux produced is of alternating nature. It is not a synchronously rotating flux, so an alternating flux acting on stationary squirrel cage rotor can not produce rotation. That is why a 1φ motor is not self starting.

Q. 2. What are the two theories related to single-phase induction motor?

Ans: These two theories are:
(a) Double revolving field theory and
(b) Cross field theory

Q. 3. What is the value of starting torque in a 1φ induction motor ?

Ans: Zero

Q. 4. How is the direction of rotation of a 1φ induction motor reversed ?

Ans: It (split phase type) is reversed by reversing the leads to the main or starting winding but not both.

Q. 5. What is the idea in double field revolving theory?

Ans: It makes use of an idea that an alternating uni-axial quantity can be represented by two oppositely rotating vectors of half magnitude.

Q. 6. What will be the slip of rotor corresponding to forward and backward rotating flux (in double revolving field theory)?

Ans: It will be S and 2 – S respectively.

Q. 7. What is done so as to make the motor self-starting?

Ans: To make the motor self-starting, it is temporarily converted into a two phase motor during starting period.

Q. 8. In split-phase machine, how can we increase the resistance of starting winding?

Ans: It may be increased either by adding a high value of resistance (R) in series with it or by choosing a high resistance fine copper wire for winding purposes.

Q. 9. In split phase motor, how many windings are used?

Ans: It has two windings namely starting winding and main winding.

Q. 10. What are the various applications of split-phase motor?

Ans: The applications are implemented in centrifugal pumps, washing machines, duplicating machines, small machine tools, etc.

Q. 11. In split-phase motors, what is the relation between starting torque and full load torque?

Ans: The starting torque is 150% to 200% of the full load torque.

Q. 12. Which type of capacitor is generally used in capacitor-start induction run motors ?

Ans: Electrolytic type

Q. 13. What will be the angle between I_s and I_m in capacitor-start induction run motors ?

Ans: It will be nearly 80°.

Q. 14. In a split-phase induction motor, how would you differentiate between main and starting windings, if the lead markings are not labelled?

Ans: It may be done by measuring the resistances of both the windings. Starting winding will have more resistance compared to the main winding.

Q. 15. Which motor is preferred for blowers and fans ?

Ans: It is split-phase motor.

Q. 16. Which motor is used for refrigeration and air conditioning?

Ans: In refrigeration and air-conditioning, 1φ, 220 V capacitor type induction motor of intermittent rating is used.

Q. 17. How will you change the direction of rotation of a capacitor start induction motor?

Ans: It may be reversed by reversing either the starting or running winding leads, not both, to the line.

Q. 18. What will happen if the starting winding of a capacitor-motor is disconnected during running condition?

Ans: The motor will continue running but will develop small torque. Once it stops, it will not start again.

Q. 19. At which places, two value capacitor run motor may be used?

Ans: It may be used where the load requirements are severe as in the case of fire strokers and compressors.

Q. 20. In a single-phase induction motor, two windings are employed. Which winding is more resistive in nature?

Ans: Auxiliary or starting winding will be more resistive in nature.

Q. 21. What will be the benefit of using a capacitor-start motor over a resistance-start split phase motor?

Ans: The capacitor-start motor will develop more starting torque compared to a resistance-start split phase motor (for same rating machines).

Q. 22. Why the angle of lag of current I_a drawn by the starting winding will be less compared to the main winding current (I_m) in split phase resistance start motor?

Ans: It is because the ratio of resistance to reactance of the auxiliary winding is more than the main winding.

Q. 23. What do you mean by shaded pole motors?

Ans: These are the motors, which use shaded poles. These motors are used in applications where the starting torque requirement of the load is low.

Q. 24. In shaded pole motors, which type of poles are used?

Ans: These motors have salient poles (on the stator).

Q. 25. What are the disadvantages of shaded pole motor?

Ans: The disadvantages include:
(1) Little overload capacity
(2) Low efficiency
(3) Low value of starting torque, etc.

Q. 26. For which devices, shaded pole motors are used?

Ans: Shaded pole motors are used for small fans, toys, ventilators, electric clocks, etc.

Q. 27. What is the role of magnetic bridges in the shaded pole motor?

Ans: The magnetic bridges cause the flux in the shading portion of the pole to lag behind the flux in the unshaded portion. Finally it gives a motion of flux across the pole face and under the influence of this moving flux a starting torque is developed.

Q. 28. Which type of motor is used in ceiling fan?

Ans: Ceiling fan is allowed to run by split-phase type 1φ induction motor.

Q. 29. For which applications, capacitor-start motors are suitable?

Ans: These are suitable for applications in pumps and compressors. These may also be used in refrigerators.

Q. 30. When the rotor of single-phase induction motor is blocked, what will be the value of slip?

Ans: The value of slip will be equal to one.

Q. 31. What are the limitations of permanent-split capacitor (PSC) motors?

Ans: The limitations of PSC motors are as follows:
(a) In these motors, paper-spaced oil filled type capacitors are used, which are very costly.
(b) Since only one capacitor is used, so a single value capacitor will provide low starting torque usually less than full load torque.

Q. 32. A split-phase motor fails to start and hums loudly. What must be the reasons?

Ans: The problem may be due to the starting winding being open or grounded or burnt out.

Q. 33. What will happen if the centrifugal switch of a two-value capacitor motor using two capacitors fails to open?

Ans: In this particular situation, electrolytic capacitor will, in all probability, suffer the breakdown.

Q. 34. Why 1φ induction motors are called as fractional kilowatt motors?

Ans: Actually single-phase motors are manufactured in fractional kilowatt range to be operated on 1-phase supply, so they are also called fractional kilowatt motors.

5

Single-Phase Transformer

Q. 1. A transformer is used to transform:

    (a) Frequency   (b) Current

    (c) Voltage   (d) Voltage and Current

Ans: (d)

Q. 2. In a transformer, open circuit test is done to obtain:

    (a) Only copper losses

    (b) Only iron losses

    (c) Both copper and iron losses

    (d) None

Ans: (b)

Q. 3. A power transformer is basically a device, called as:

    (a) Constant voltage device

    (b) Constant main flux device

    (c) Constant current device

    (d) Constant power device

Ans: (b)

Q. 4. A step-up transformer is one, which increases:

    (a) Current   (b) Voltage

    (c) Power   (d) Both voltage and current

Ans: (b)

Q. 5. The main purpose of using core in a transformer is to:

    (a) Decrease reluctance of common magnetic circuit

    (b) Prevent hysteresis losses

(c) Decrease core losses

(d) Reduce eddy current losses

Ans: **(a)**

Q. 6. EMF equation of transformer depends upon (which factors):

(a) Maximum value of flux in the core

(b) Number of turns in the winding

(c) Frequency

(d) All of the above

Ans: **(d)**

Q. 7. The value of flux in the electromotive force equation of a single-phase transformer is:

(a) Average value of flux

(b) RMS value of flux

(c) Maximum value of flux

(d) None of the above

Ans: **(c)**

Q. 8. In an AC excited iron core of transformer, maximum flux created can be obtained by:

(a) Both current and frequency

(b) Voltage applied only

(c) Core reluctance

(d) Both impressed voltage and frequency

Ans: **(d)**

Q. 9. In a transformer, high frequency humming noise is produced due to:

(a) Magnetostriction

(b) Coolant used

(c) Core

(d) Laminations (being not sufficiently tight)

Ans: **(a)**

Q. 10. The thickness of lamination of a 50 Hz transformer is:
(a) 2 mm (b) 3 mm
(c) 0.35 mm (d) 3.5 mm
Ans: **(c)**

Q. 11. The rating of a transformer is given in:
(a) kWh (b) kVA
(c) kV (d) kA
Ans: **(b)**

Q. 12. In a transformer, the phasor of flux:
(a) Leads induced emf by 180°
(b) Lags the induced emf by 90°
(c) Leads induced emf by 90°
(d) Lags induced emf by 180°
Ans: **(c)**

Q. 13. In transformer, energy transformation from one winding to another winding takes place:
(a) Magnetically (b) Mutually
(c) Electromagnetically (d) Electrically
Ans: **(a)**

Q. 14. The purpose of transformer oil is to provide:
(a) Insulation only
(b) Lubrication only
(c) Cooling only
(d) Cooling and insulation both
Ans: **(d)**

Q. 15. In an ordinary two winding transformer, the electromotive force per turn in secondary winding is always .... the induced emf per turn in primary:
(a) Equal to (b) Less than
(c) Equal to $1/k$ times (d) Equal to k times
Ans: **(a)**

Q. 16. Transformer stampings are varnished before being used to build the core to:

(a) Reduce iron loss

(b) Reduce eddy current loss

(c) Reduce cu loss

(d) Increase air gap between stampings

Ans: (b)

Q. 17. When transformer is supplied with input power at rated voltage and no load condition, then it mainly consists of:

(a) Core loss   (b) Both copper and core loss

(c) Eddy current loss   (d) Hysteresis loss

Ans: (a)

Q. 18. Minor insulation of a transformer is the insulation between:

(a) Layers of windings only

(b) Turns of the windings and also between layers of the windings

(c) Coils only

(d) Two segments of high voltage winding only

Ans: (b)

Q. 19. The application of transformer includes:

(a) As a coupling device

(b) To measure voltage and current

(c) As impedance-matching device for maximum power transfer in low power electronic and control circuits

(d) All of the above

Ans: (d)

Q. 20 In case of large size transformers, best utilization of available core space can be made by using:

(a) Stepped core section

(b) Rectangular cross-section

(c) Either square or rectangular section

(d) Square cross-section

Ans: (a)

Q. 21. A transformer may not change the level (raise or lower) of DC supply, because:

(a) A DC circuit has more losses

(b) A DC circuit is less efficient compared to AC circuit

(c) Faraday's law of electromagnetic induction is not valid since the rate of change of flux is zero

(d) A DC circuit consumes more power

Ans: (c)

Q. 22. The flux in the core of a practical transformer, with a resistive load:

(a) Decreases with increased load

(b) Increases as the square root of load

(c) Increases linearly with load

(d) Remains constant irrespective of load conditions

Ans: (d)

Q. 23. The value of flux in the emf equation for a transformer is:

(a) RMS value   (b) Average value

(c) Maximum value   (d) Instantaneous value

Ans: (c)

Q. 24. The iron core in a transformer provides a ….. path of the main flux:

(a) Low resistance   (b) High reluctance

(c) Low reluctance   (d) High conductivity

Ans: (c)

Q. 25. In a transformer, the voltage transformation ratio is:

(a) N_2/N_1 (b) E_2/E_1

(c) I_1/I_2 (d) All of the above

Ans: (d)

Q. 26. The no load current in a transformer lags the applied voltage by:

(a) 75°   (b) 90°

(c) 180°   (d) 0°

Ans: (a)

Q. 27. The no load current is nearly .... of full load current:

(a) 1% to 3%   (b) 6% to 9%

(c) 3% to 6%   (d) 2% to 3%

Ans: (a)

Q. 28. The leakage flux depends upon:

(a) Frequency   (b) Value of load current

(c) Mutual flux   (d) Maximum flux in the core

Ans: (b)

Q. 29. The value of inductive reactance of a transformer depends upon:

(a) Magnetic flux   (b) Leakage flux

(c) Magnetomotive force   (d) Both (a) and (b)

Ans: (b)

Q. 30. The leakage flux is the flux, which links with:

(a) Only primary winding

(b) Only secondary winding

(c) Either primary or secondary winding

(d) Both (a) and (b)

Ans: (c)

Q. 31. Silicon steel is normally preferred for core of transformer, because it:

(a) Helps to reduce leakage flux

(b) Helps to reduce core losses

(c) Helps to reduce copper losses

(d) Helps to increase tensile strength

Ans: (b)

Q. 32. The oil, used in transformer must be free from moisture contents because:

(a) It increases emf in the secondary winding

(b) It reduces the dielectric strength of oil

(c) It increases core losses

(d) It increases the oil density

Ans: **(b)**

Q. 33. The function of conservator is to:

(a) Protect the transformer from damage, when oil expands due to heating

(b) Provide safety from external effects

(c) Supply cooling oil to transformer in the time of need

(d) Raise the emf in the secondary winding

Ans: **(a)**

Q. 34. The primary and secondary values of induced emf, in case of a single-phase transformer are:

(a) 180° (out of phase)

(b) 90° (out of phase)

(c) In phase with each other

(d) 45° (out of phase)

Ans: **(c)**

Q. 35. When the transformer is loaded, then the mutual flux may be changed by changing:

(a) Secondary current

(b) Load impedance

(c) Load impedance and secondary current

(d) Reluctance of magnetic path

Ans: **(d)**

Q. 36. Under the no load condition, applied voltage is approximately balanced by:

(a) Secondary induced emf (E_2)

(b) Primary induced emf (E_1)

(c) Terminal voltage across load (V_2)

(d) None of these

Ans: **(b)**

Q. 37. The value of cosθ, at which the transformer operates is:

(a) Unity

(b) 0.85 leading

(c) 0.85 lagging

(d) Depends upon the power factor of load

Ans: **(d)**

Q. 38. Sludging of transformer oil means:

(a) Evaporation of oil due to heating

(b) Formation of dust at the bottom of transformer tank

(c) Freezing of oil in winter season

(d) Formation of semi-solid hydrocarbon due to heat and oxidation

Ans: **(d)**

Q. 39. A transformer may have two or more ratings depending upon:

(a) The type of core used

(b) The type of insulation used

(c) The type of cooling used

(d) The type of winding used

Ans: **(c)**

Q. 40. In a transformer, the core loss is represented by:

(a) Shunt resistance

(b) Shunt inductance

(c) Series inductance

(d) Series resistance

Ans: **(a)**

Q. 41. **The purpose of performing open circuit test (OC test) on a transformer is to measure:**

(a) Core loss (iron loss)　　(b) Copper loss

(c) Total loss　　　　　　　(d) Eddy current loss

Ans: **(a)**

Q. 42. **The rating of a transformer is given in kVA, but not in kW, because:**

(a) Total transformer loss depends upon volt-ampere (*VI*)

(b) kVA is a fixed value but kW depends on load power factor

(c) It makes the transformer more suitable for external uses

(d) It has become customary

Ans: **(a)**

Q. 43. **In distribution transformers, core losses are:**

(a) Negligible compared to full load copper losses

(b) Less than full load copper losses

(c) Equal to full load copper losses

(d) More than full load copper losses

Ans: **(b)**

Q. 44. **The short circuit test (SC test) is done on a 1φ transformer to obtain:**

(a) Copper loss　　　　　　(b) Core loss (iron loss)

(c) Total amount of loss　　(d) None of the above

Ans: **(a)**

Q. 45. **Which one of the followings is a common method for cooling a power transformer?**

(a) Oil cooling

(b) Air blast cooling

(c) Oil natural air natural cooling

(d) None of the above

Ans: **(a)**

Q. 46. When the supply frequency is increased, in case of a single-phase transformer, the rating of transformer:

(a) Increases

(b) Decreases

(c) May increase in a particular condition

(d) Neither increases nor decreases

Ans: **(a)**

Q. 47. When a 400 Hz transformer is operated at 50 Hz, its kVA rating is:

(a) Increased 16 times

(b) Unaffected

(c) Reduced to 1/8th

(d) Reduced to 1/16th

Ans: **(c)**

Q. 48. When the short circuit test is performed, then usually:

(a) High voltage winding is short circuited

(b) Low voltage winding is short circuited

(c) Any side may be short circuited

(d) None of the above

Ans: **(b)**

Q. 49. In performing open circuit test (OC test),..... is applied

(a) Rated primary current

(b) Rated frequency

(c) Rated transformer voltage

(d) Direct current

Ans: **(c)**

Q. 50. In case of concentric windings (used in transformer), low voltage windings are kept near the core, because:

(a) It increases the tensile strength of arrangement

(b) It reduces iron loss (core loss)

(c) It reduces leakage flux

(d) It reduces the insulation requirement

Ans: **(d)**

Q. 51. While performing the open circuit and short circuit tests on a transformer to determine parameters, the status of the low voltage (LV) and high voltage (HV) windings will be such that:

(a) In OC test, HV is open and in SC, HV is shorted

(b) In OC test, LV is open and in SC, LV is shorted

(c) In OC test, HV is open and in SC, LV is shorted

(d) In OC test, LV is open and in SC, HV is shorted

Ans: (c)

Q. 52. When rated DC voltage is applied instead of AC supply to transformer, then:

(a) Primary winding of transformer will burn

(b) Flux produced will be less

(c) Secondary winding of transformer will burn

(d) Self induced emf will be less compared to mutually induced emf

Ans: (a)

Q. 53. The magnetizing component of no load current (I_m or I_μ):

(a) Lags behind the applied voltage by 90°

(b) Leads the applied voltage by 90°

(c) Out of phase with applied voltage

(d) In phase with mutually induced emf

Ans: (a)

Q. 54. When the transformer supplies power to a pure resistive (unity power factor) load, then power factor on primary side will be:

(a) Nearly unity

(b) Zero

(c) About 0.95 (lagging)

(d) About 0.95 (leading)

Ans: (c)

Q. 55. The voltage applied to the HV side of a transformer during short circuit test is 2% of its rated voltage. The core loss will be, how much percentage of rated core loss?

(a) 0.04   (b) 0.4

(c) 0.004   (d) 4

Ans: **(a)**

Q. 56. The mutual flux in transformer is:

(a) Constant at all loaded conditions

(b) Low at all loaded conditions

(c) High at all loaded conditions

(d) High at low load and low at high load condition

Ans: **(a)**

Q. 57. The transformer oil has to fulfill certain specifications, which include:

(a) Low viscosity

(b) Good resistance to emulsion

(c) High flash point and dielectric strength

(d) All of the above

Ans: **(d)**

Q. 58. The functions of insulating oil are:

(a) It provides additional insulation

(b) It protects the insulation from dirt and moisture

(c) It carries away the heat generated in the cores and coils

(d) All of the above

Ans: **(d)**

Q. 59. Which one of the followings are the types of concentric winding?

(a) Helical winding   (b) Continuous disc winding

(d) Crossover winding   (d) All of the above

Ans: **(d)**

Q. 60. The higher content of silicon in the core:

(a) Increases the resistivity of the core

(b) Reduces the eddy current loss

(c) Increases the hysteresis loss

(d) Both (a) and (b)

Ans: **(d)**

Q. 61. The steel used in transformer gets brittle, if the silicon content is increased beyond:

(a) 2%   (b) 2.5%

(c) 2.75%   (d) 5%

Ans: **(d)**

Q. 62. Which of the following is not considered as a method to reduce leakage flux in a transformer?

(a) Increasing height of the window

(b) Sandwitching primary and secondary windings

(c) Adopting shell type construction and arranging both the windings concentrically

(d) Decreasing height of the window

Ans: **(d)**

Q. 63. When the secondary winding resistance (R_2), is transferred from secondary winding to primary winding, then it is named as:

(a) R_2'

(b) R_2''

(c) R_2/K^2

(d) Both (a) and (c)

Ans: **(d)**

Q. 64. When primary winding resistance is transferred to secondary side, then it is called as:

(a) R_1' (b) $K^2 R_1$

(c) R_1/K^2 (d) Both (a) and (b)

Ans: **(d)**

Q. 65. In a transformer, the total resistance as referred to primary is called as:

(a) R_{02} (b) R_{01}

(c) $R_1 + R_2'$ (d) Both (b) and (c)

Ans: **(d)**

Q. 66. The equivalent resistance of the primary of a transformer having $K = 5$ and $R_1 = 0.1$ ohms, when referred to secondary becomes:

(a) 0.5 ohms   (b) 2.5 ohms

(c) 0.25 ohms   (d) 25 ohms

Ans: **(b)**

Q. 67. Which of the following tests, is not performed in case of single-phase transformer?

(a) Voltage ratio test

(b) Polarity test

(c) Sumpner's (back to back) test

(d) Hopkinson's test

Ans: **(d)**

Q. 68. When the supply frequency is tripled in case of a transformer, then:

(a) Hysteresis loss becomes triple

(b) Eddy current loss becomes 27 times

(c) Iron loss becomes double

(d) Copper loss becomes triple

Ans: **(a)**

Q. 69. Which of the following tests, give the summation of eddy current and hysteresis losses in a single-phase transformer?

(a) Core loss (Iron loss) test

(b) Open circuit test

(c) Polarity test

(d) Back to back test

Ans: **(b)**

Q. 70. In a 400 V, 50 Hz transformer, the total iron loss is 2500 watts. When applied voltage and frequency are reduced to 200 V and 25 Hz respectively, the corresponding loss becomes 850 watts. What is the value of eddy current loss at normal voltage and frequency?

(a) 1600 watts

(b) 1200 watts

(c) 1500 watts

(d) 1250 watts

Ans: (a)

Q. 71. In a transformer, the core loss is 52 watts at 40 Hz and 90 watts at 60 Hz, measured at same peak flux density. What will be the value of eddy current and hysteresis losses at 50 Hz, respectively?

(a) 25 W and 45 W

(b) 45 W and 25 W

(c) 30 W and 45 W

(d) 45 W and 30 W

Ans: (a)

Q. 72. If P_c and P_{sc} represent core and full load ohmic losses respectively, the maximum kVA delivered to load corresponding to maximum efficiency is equal to rated kVA multiplied by:

(a) $(P_c/P_{sc})^2$ (b) $(P_{sc}/P_c)^2$

(c) P_c/P_{sc} (d) $\sqrt{(P_c/P_{sc})}$

Ans: (d)

Q. 73. The efficiency of a power transformer may be easily obtained by:

(a) Open circuit test

(b) Polarity test

(c) Open and short circuit tests

(d) Regenerative test

Ans: (c)

Q. 74. Which of the following is the necessary condition for obtaining maximum efficiency of a transformer?

    (a) Copper loss > Iron loss

    (b) Copper loss < Iron loss

    (c) Copper loss = Core loss

    (d) None of the above

Ans: (c)

Q. 75. If P_1 and P_2 are iron and copper losses of a transformer on full load respectively, then what will be the ratio of P_1 and P_2, such that maximum efficiency occurs at 75% of full load?

    (a) $P_1/P_2 = 9/16$ (b) $P_1/P_2 = 16/9$

    (c) $P_1/P_2 = 4/3$ (d) $P_1/P_2 = 3/4$

Ans: (a)

Q. 76. At which value of power factor, efficiency of transformer will be maximum?

    (a) 0.85 (lagging)   (b) 0.85 (leading)

    (c) 0.90 (leading)   (d) Unity

Ans: (d)

Q. 77. If the transformer efficiency is 95% at 0.85 power factor (lagging), then what will be the value of efficiency at 0.85 power factor (leading)?

    (a) 95%   (b) More than 95%

    (c) Less than 95%   (d) None of the above

Ans: (a)

Q. 78. The efficiency of a single-phase transformer, whose rating is 100 kVA is 0.98 (98%) at full load as well as at half load conditions. What will be the full load copper loss?

    (a) Less than core loss

    (b) Equal to core loss

    (c) More than core loss

    (d) None of the above

Ans: (c)

Q. 79. The rating of a single-phase transformer is 200 kVA and voltage ratio is 2400 V/240 V. It has core loss of 1.8 kW at rated voltage. The equivalent resistance is 1.1%. What will be the value of transformer efficiency at a PF of 0.9 (at full load condition)?

(a) 97.80%

(b) 97.82%

(c) 97.08%

(d) 97.85%

Ans: **(b)**

Q. 80. In a 1φ transformer, iron losses and copper losses are 40.5 kW and 50 kW respectively. At what fraction of load, will the efficiency be maximum?

(a) 0.9   (b) 0.92

(c) 0.57   (d) 0.50

Ans: **(a)**

Q. 81. In which range, the efficiency of a transformer will lie?

(a) 90% to 99%

(b) 92% to 98%

(c) 96% to 97%

(d) 96% to 99%

Ans: **(d)**

Q. 82. Let P_{Fe} = iron loss and P_{Cu} = copper loss (Both at full load condition). What will be the ratio of P_{Fe}/P_{Cu}, such that the maximum efficiency occurs at 75% of full load?

(a) 0.5625   (b) 0.4625

(c) 0.6025   (d) 0.5629

Ans: **(a)**

Q. 83. In a 100 kVA transformer, iron losses and copper losses are 900 kW and 1600 kW respectively. The maximum efficiency will occur at:

(a) 75   (b) 75.75

(c) 90.02   (d) 75.5

Ans: **(a)**

Q. 84. In case of a single-phase transformer, data is as follows:
Rating = 100 kVA, Iron loss = 600 watts
Cu loss = 1.5 kW .... (Both losses at full load current)
What will be the value of transformer efficiency at
100 kVA output at 0.8 PF (lagging)?

(a) 98.09%   (b) 97.44%

(c) 98.00%   (d) 97.47%

Ans: (b)

Q. 85. A curve is drawn between full load and efficiency of a
single-phase transformer. At which PF, the curve will
attain maximum height?

(a) 0.6 PF

(b) 0.8 PF

(c) 0.9 PF

(d) 1.0 PF

Ans: (d)

Q. 86. The regulation of a transformer is defined as:

(a) Change in flux from no load to full load

(b) Fall in terminal voltage under loaded condition

(c) Change in secondary terminal voltage from no load
to full load as a percentage of secondary no load
terminal voltage

(d) None of the above

Ans: (c)

Q. 87. What will be the value of regulation of transformer,
when heat dissipation losses are 1% of output and
reactance drop is 5% of the voltage, at a PF of 0.80
(leading)?

(a) –2.2%

(b) 1%

(c) 3.8%

(d) 3.27%

Ans: (a)

Q. 88. In a single-phase transformer, data is as follows:
Rating of transformer = 40 kVA,
Voltage ratio = 6600/ 250,
$R_1 = 10 \, \Omega$, $R_2 = 0.02 \, \Omega$, X_l (referred to primary) = 35 Ω
What will be the full load regulation at 0.8 (lagging PF)?

(a) 3.7%   (b) 2.2%
(c) 2.0%   (d) 3.5%

Ans: **(a)**

Q. 89. In a single-phase transformer, the voltage regulation at full load condition and at a power factor of 0.85 (lagging) is 5%. What will be the value of voltage regulation at full load and at 0.85 PF (leading)?

(a) May be positive
(b) May be negative
(c) Remains constant, irrespective of PF
(d) Reduces and may even become negative

Ans: **(d)**

Q. 90. In a large transformer, voltage regulation is influenced by:

(a) I_0 and $\cos\phi_L$ (b) $R_{windings}$ and $\cos\phi_L$
(c) ϕ_l and $\cos\phi_l$ (d) $R_{windings}$ and core loss

Ans: **(b)**

Q. 91. In a power transformer, full load voltage regulation becomes zero, when load power factor is approximately:

(a) Unity and leading   (b) Unity and lagging
(c) Zero and leading   (d) 0.5 and lagging

Ans: **(a)**

Q. 92. At full load with 0.8 PF (lagging), voltage regulation of a transformer is 3%. What will be its value at unity power factor?

(a) 3%   (b) Negative
(c) Always positive   (d) 2.75%

Ans: **(b)**

Q. 93. What will be the voltage regulation of a transformer, when ohmic loss is 1% of output and reactance drop is 5% of voltage, at unity power factor?

(a) 1%   (b) 2%
(c) 1.5%   (d) 3%

Ans: (a)

Q. 94. A transformer has a percentage resistance of 1% and percentage reactance of 4%. Its regulation at power factor 0.8 lagging and 0.8 leading are respectively?

(a) 3.2% and –1.6%   (b) 6% and –4%
(c) 2% and –3%   (d) None of these

Ans: (a)

Q. 95. All day efficiency of a transformer depends primarily upon:

(a) Its copper loss   (b) The duration of load
(c) Amount of load   (d) Both (b) and (c)

Ans: (d)

Q. 96. Which of the following statements is wrong?

(a) Auto transformer uses less copper compared to ordinary two winding transformer
(b) It is cheaper
(c) It is used where transformation ratio differs little from unity
(d) In this autotransformer, primary and secondary windings are isolated

Ans: (d)

Q. 97. The uses of autotransformer include:

(a) As interconnecting transformers in 132 kV/330 kV system
(b) As furnace transformers
(c) As autostarters in case of induction motors
(d) All of the above

Ans: (d)

Q. 98. In autotransformer, which of the relations is true?

    (a) VA (auto) > VA (two winding)

    (b) VA (auto) < VA (two winding)

    (c) VA (auto) = VA (two winding)

    (d) None of the above

Ans: **(a)**

Q. 99. Two transformers connected in parallel share load in the ratio of their kVA ratings only if their per-unit impedances (on their own kVA's) are:

    (a) Purely resistive

    (b) Equal

    (c) In the inverse ratio of their ratings

    (d) Purely reactive

Ans: **(b)**

Q. 100. A 2500 V/250 V, 25 kVA transformer is connected as an autotransformer to give 2300 V/ 2750 V. What will be its rating?

    (a) 225 kVA hence 225 kVA is transferred conductively

    (b) 250 kVA hence 250 kVA is transferred inductively

    (c) 275 kVA hence 250 kVA is transformed conductively

    (d) 250 kVA hence 250 kVA is transformed inductively

Ans: **(c)**

MISCELLANEOUS

Q. 1. In an ideal transformer, which statement is not correct?

    (a) No iron loss

    (b) Zero magnetizing current

    (c) Very small winding resistance

    (d) No magnetic leakage

Ans: **(c)**

Q. 2. Which of the following relations are true?

(a) P_h is proportional to $f(\varphi_{max})^x$, where x lies between 1.5 and 2.5

(b) P_e is proportional to $f^2(\varphi_{max})^2$, where x lies between 1.5 and 2.5

(c) $P_{core} = P_h + P_e$

(d) All of the above

Ans: (d)

Q. 3. On open circuit condition, a 2200 V/250 V transformer takes 0.5 A at a PF of 0.3. What will be the value of active component of no load primary current?

(a) 0.50 A   (b) 0.45 A

(c) 0.15 A   (d) 0.17 A

Ans: (c)

Q. 4. In a single-phase transformer, I_2' is called as:

(a) Load component of primary current

(b) Primary component of secondary current

(c) Active component of primary current

(d) Reactive component of primary current

Ans: (a)

Q. 5. Which of the following relations is not correct?

(a) Leakage fluxes are practically proportional to the currents in the respective windings

(b) Leakage flux links one or the other winding but not both

(c) Main useful flux φ decreases slightly with increase in load

(d) Induced emf in secondary winding (E_2) meets the resistive and reactive drops in the primary winding

Ans: (d)

Q. 6. Which of the following relations is correct?

(a) $R_{02} = K^2 R_{01}$ (b) $R_{01} = R_1 + R_2/K^2$

(c) $Z_{02} = K^2 Z_{01}$ (d) All of the above

Ans: (d)

Q. 7. The recommended method for measuring transformer winding resistance is:

(a) Ammeter-voltmeter method

(b) Wheatstone bridge method

(c) Either (a) or (b)

(d) Kelvin's double bridge method

Ans: (c)

Q. 8. All day efficiency of transformer may also be called as:

(a) Operational efficiency   (b) Energy efficiency

(c) None   (d) Both (a) and (b)

Ans: (d)

Q. 9. Transformer core is built from thin stampings, because:

(a) It reduces the tensile strength

(b) It reduces core loss

(c) It reduces hysteresis loss

(d) It reduces eddy current loss

Ans: (d)

Q. 10. In ferromagnetic materials, magnetostriction noise developes due to:

(a) Cu loss

(b) Core loss

(c) Changes in dimensions of crystals under AC excitation

(d) Eddy current loss

Ans: (c)

Q. 11. Depending upon the type of construction used, the transformers are classified into following categories?

(a) Only core type

(b) Only shell type

(c) Spiral core or wound core type

(d) All of the above

Ans: (d)

Q. 12. **Whenever a sudden short circuit fault occurs, on the line fed by a transformer, the mechanical stresses will be maximum at:**
(a) Coil ends   (b) Yoke
(c) Core ends   (d) Coils
Ans: **(a)**

Q. 13. **Which type of cooling methods provide the best cooling?**
(a) ONWF   (b) Open type dry
(c) OFWF   (d) ONAF
Ans: **(c)**

Q. 14. **The magnetizing flux development, in case of transformer is due to:**
(a) Primary induced emf only
(b) Applied primary voltage
(c) Secondary mmf
(d) Total mmf
Ans: **(b)**

Q. 15. **A single-phase transformer is supplying a load, whose nature is lagging. What will be the effect on the voltage regulation of transformer, if a capacitor is connected in parallel to the load?**
(a) Regulation will increase
(b) Regulation will decrease
(c) Regulation will remain constant
(d) None of the above
Ans: **(b)**

Q. 16. **When an autotransformer is made using a two winding transformer, then its efficiency at full load will:**
(a) Increase   (b) Decrease
(c) Remain constant   (d) Become 90%
Ans: **(a)**

Q. 17. In a 500 kVA transformer, efficiency is 95% at full load and 60% of full load, both at unity power factor. What will be the values of P_c and P_i ?

(a) P_c = 16.45 kW and P_i = 9.87 kW

(b) P_c = 7 kW and P_i = 9.85 kW

(c) P_c = 16.45 kW and P_i = 9.27 kW

(d) P_c = 16.45 kW and P_i = 9.29 kW

Ans: (a)

Q. 18. A ferrite core consists of lower specific eddy current loss compared to an iron core, because ferrite core consists of:

(a) High permeability   (b) High electrical resistance

(c) Low conductivity   (d) High value of reluctance

Ans: (b)

Q. 19. In transformer primary and secondary windings are interlaced for the purpose of:

(a) Reduction of core loss

(b) Increment in eddy current heating

(c) Reduction in cost

(d) Reduction of leakage reactance

Ans: (d)

Q. 20. The purpose of stepped core in case of transformer is to reduce:

(a) Cost

(b) Volume of iron

(c) The number of limbs

(d) Copper volume

Ans: (d)

Q. 21. When load is allowed to vary in secondary side, which value will follow the variation in load?

(a) Core losses   (b) I_m

(c) I_w (d) Cu losses

Ans: (d)

Q. 22. Due to the addition of tubes, in a transformer tank, heat dissipation capacity improves, because:

(a) Dissipation takes place both by conduction and convection

(b) Dissipation takes place both by convection and radiation

(c) Additional dissipation by radiation only

(d) Additional cooling surface provided

Ans: **(d)**

Q. 23. To obtain minimum inrush current, a single-phase transformer is to be switched on to supply. The moment at which, the switch must be closed, will be at:

(a) 0.5 × maximum supply voltage

(b) 0.707 × maximum voltage

(c) Maximum supply voltage

(d) 0.55 × maximum supply voltage

Ans: **(c)**

Q. 24. A 20 kVA 2000 V/200 V transformer has 8% leakage impedance. How much voltage must be applied on the HV side, so that full load current may circulate (with LV winding shorted)?

(a) 160 V

(b) 150 V

(c) 145.5 V

(d) 120 V

Ans: **(a)**

Q. 25. Voltage regulation of transformer is a measure of:

(a) Voltage

(b) Current amplitude

(c) Voltage drop due to loading

(d) Frequency variation

Ans: **(c)**

Q. 26. The percentage regulation is defined as:

    (a) (Terminal voltage on no load – Terminal voltage on load)/Terminal voltage on no load × 100

    (b) Voltage drop in transformer at load/No load rated voltage (secondary) × 100

    (c) $I_2R_{02}\cos\phi - I_2X_{02}\sin\phi$/No load rated secondary voltage × 100

    (d) All of the above

Ans: (d)

Q. 27. Which will improve the mutual coupling between two circuits of a transformer?

    (a) Low reluctance magnetic core

    (b) Smaller cross-section

    (c) Lesser number of turns in secondary

    (d) Less magnetic material

Ans: (a)

Q. 28. When two identical transformers are loaded, then the efficiency of transformer can be obtained by:

    (a) Voltage ratio test   (b) Back to back test

    (c) Open circuit test   (d) Polarity test

Ans: (b)

Q. 29. At which kVA rating of transformer, conservator tank is provided?

    (a) 50 kVA or above

    (b) More than 100 kVA

    (c) More than 20 kVA

    (d) 100 kVA to 200 kVA

Ans: (a)

Q. 30. What is the usual secondary voltage rating of a PT?

    (a) 150 V   (b) 110 V

    (c) 1500 V   (d) 50000 V

Ans: (b)

Q. 31. When the transformer oil is new, it appears as a:

(a) Clear light colour with a faint characteristic smell

(b) Clear yellow colour with a faint characteristic smell

(c) Brownish colour

(d) Pink colour

Ans: **(a)**

Q. 32. A single-phase transformer is used to step down the voltage from 220 V to 110 V. However, if an autotransformer is used for the same purpose, then what will be the ratio of copper weights in them?

(a) 0.25   (b) 0.75

(c) 0.55   (d) 0.5

Ans: **(d)**

Q. 33. It is actually not difficult to insulate transformers for very high voltages, because:

(a) It has no teeth

(b) The windings can be immersed in oil

(c) It has no rotating part

(d) All of the above

Ans: **(d)**

Q. 34. If in a single-phase transformer applied voltage is increased by 50% and frequency is reduced to 50%, then what will be the maximum core flux density?

(a) Becomes 3 times

(b) Becomes 1/2 times

(c) Becomes 2 times

(d) Remains constant

Ans: **(a)**

Q. 35. The magnetizing component of no load current (I_0) in a transformer is rich in:

(a) 3rd harmonic   (b) 7th harmonic

(c) 12th harmonic   (d) 5th harmonic

Ans: **(a)**

Q. 36. In a transformer, if $X_l = 2R_1 = 2R_2$ and secondary winding is short circuited, then the input power factor may be:

(a) $\dfrac{1}{\sqrt{2}}$

(b) $\dfrac{1}{\sqrt{3}}$

(c) $\dfrac{2}{\sqrt{5}}$

(d) $\dfrac{1}{\sqrt{5}}$

Ans: (c)

Q. 37. Which one of the following losses does not occur in a single-phase transformer?

(a) Core loss (iron loss) or P_i

(b) Copper loss (I^2R loss) or P_c

(c) Dielectric loss

(d) Friction loss

Ans: (d)

Q. 38. A load of 5 kW at 115 V and unity power factor is supplied by a autotransformer. If the primary voltage is 230 V, then determine the value of power conducted?

(a) 2.5 kW

(b) 7.0 kW

(c) 7.5 kW

(d) 3.9 kW

Ans: (a)

Q. 39. To prevent maloperation of differentially connected relay, while energizing a transformer, the relay restraining coil is based with:

(a) 2nd harmonic current

(b) 5th harmonic current

(c) 7th harmonic current

(d) 9th harmonic current

Ans: (a)

Q. 40. The short circuit voltage of a transformer mainly depends on the:

(a) Magnitude of leakage flux

(b) Leakage reactance of secondary winding

(c) Leakage reactance of primary winding

(d) None of these

Ans: (a)

Q. 41. Which one of the following statements regarding a transformer is wrong?

(a) In a shell type transformer, the primary and secondary windings are wound on separate limbs

(b) The open circuit test is performed to obtain core loss

(c) The short circuit test is performed to obtain Cu loss

(d) In a core type transformer, the primary and secondary windings are wound on separate limbs

Ans: (a)

Q. 42. A transformer, when supplying a load, maintained 11 kV across load terminals. When the load was switched off, the terminal voltage became 11550 V. What is the voltage regulation at this load?

(a) 11.55%   (b) 5.5%

(c) 5%   (d) 55%

Ans: (c)

Q. 43. The marked increase in kVA capacity produced by connecting a 2 winding transformer as an autotransformer is due to:

(a) Establishment of conductive link between primary and secondary

(b) Increase in the efficiency

(c) Reduction in leakage flux

(d) Decrement in secondary voltage

Ans: (a)

Q. 44. The conditions for satisfactory parallel operation of single-phase transformers are:

(a) Ratio of winding resistances to reactance for the transformers should be equal

(b) Per unit percentage impedance of two transformers must be equal

(c) The turn ratios of the transformers must be equal

(d) All of the above

Ans: (d)

Q. 45. In case of a single-phase transformer, which test must be carried out by mutual agreement between the purchaser and the supplier?

(a) Dielectric test

(b) Measurement of power taken by fans and oil pumps

(c) Measurement of harmonics of no load current

(d) All of the above

Ans: (d)

FILL IN THE BLANKS

1. Transformer is an AC machine that transfers .... energy from one electric circuit to another circuit, without change of frequency.

Ans: Electrical

2. It is also called as static transformer, because its construction consists of no .... parts.

Ans: Moving (Rotating)

3. Transformer efficiency lies between ....

Ans: 96% to 99%

4. The core is made up of thin soft .... or silicon steel laminations to provide a path of .... reluctance to magnetic flux.

Ans: Iron; Low

5. In transformers, the supply voltage is of .... nature.

Ans: Alternating

6. The cores are built up of sheet iron or alloyed steel so as to keep down the .... loss to a minimum.

Ans: Eddy current loss

7. In case of transformer, joints in alternate layers are staggered to avoid the presence of narrow.......... right through the cross-section of the core.

Ans: Air gap

8. In actual transformer, primary and secondary coils are arranged on each limb so as to keep down ....

Ans: **Magnetic leakage**

9. No load current (I_0) has a magnetizing component (I_m), which lags ....behind V_1.

Ans: **90°**

10. In a transformer, working component (I_w) produces the necessary real power to supply.......... in the iron.

Ans: **Hysteresis and Eddy current losses**

11. In a transformer, the magnitude of emf induced depends upon their number of ....

Ans: **Turns**

12. Magnitude of induced emf in primary winding remains approximately equal (slightly less) but opposite to ....

Ans: **Applied voltage**

13. No load current (I_0) may be broken up into two components .... and ....

Ans: I_m **and** I_e

14. A transformer must never be connected to a .... source.

Ans: **DC**

15. The operating principle of a transformer is based on .... induction.

Ans: **Mutual**

16. In a step down transformer, secondary turns are .... than primary turns.

Ans: **Less**

17. In an ideal transformer on no load, the primary voltage is balanced by ....

Ans: **Primary induced emf**

18. Flux (φ), which is common to both primary and secondary windings is taken as the .... axis.

Ans: **Reference**

19. When the secondary side of transformer is connected to load, flux produced by secondary current will also complete their path through the ....

Ans: **Air**

20. The flux, linking individual windings in a transformer is called .... flux.

Ans: **Leakage**

21. The two emfs induced due to the main flux and leakage flux, can be thought to be equivalent to voltage drops due to two reactances, namely the voltage drop due to .... reactance and .... reactance.

Ans: **Magnetizing; Leakage**

22. Leakage fluxes are proportional to the .... flowing through the respective windings.

Ans: **Currents**

23. The effect of leakage fluxes may be considered as equivalent to two reactors X_1 and X_2 connected in .... with primary and secondary windings respectively, causing voltage drops.

Ans: **Series**

24. The no load current (I_0) drawn by a transformer is about .... of the rated full load current.

Ans: **3% to 5%**

25. Voltage regulation of a transformer is defined as the change in secondary terminal voltage from ....

Ans: **No load to full load**

26. The losses in a transformer can be estimated by performing two tests, namely ....

Ans: **Open circuit test and short circuit test**

27. In short circuit test, one winding usually .... winding is short circuited through an ammeter.

Ans: **Low voltage**

28. Core loss (iron loss) is composed of hysteresis loss and .... losses.

Ans: **Eddy current**

29. I^2R loss at no load remains nearly about .... (in percentage) of the full load copper loss.

Ans: **0.09% to 0.25%**

30. The mmf produced by the secondary winding will .... the primary mmf.

Ans: **Oppose**

31. Copper losses in a power transformer are proportional to the square of the .... current.

Ans: **Load**

32. The short circuit test in a transformer is used to obtain the .... loss at full load.

Ans: **Copper**

33. Silicon steel, which is used in transformer cores, contains .... percent of silicon.

Ans: **Four**

34. In open circuit test, the power consumed by the transformer at no load can be taken as the .... loss of the transformer.

Ans: **Core**

35. In short circuit test of a transformer, usually the low voltage side is short circuited by a .... conductor.

Ans: **Thick**

36. Back to back test on transformers is also called the ....

Ans: **Sumpner's test or regenerative test**

37. To determine the maximum temperature rise, it becomes necessary to conduct a .... test on a transformer.

Ans: **Full load**

38. Back to back test is performed on two .... single-phase transformers.

Ans: **Identical**

39. The transformer efficiency is the ratio of output power to .... power.

Ans: **Input**

40. In case of transformer, the efficiency for a given power factor is maximum, when the variable copper loss is .... to the constant iron (core) loss.

Ans: Equal

41. The highest possible efficiency occurs at .... power factor.

Ans: Unity

42. Equivalent resistance in per unit is .... to the full load copper loss in per unit.

Ans: Equal

43. For a constant load current, maximum efficiency occurs, when the load power factor is ....

Ans: Unity

44. Energy efficiency of a transformer is defined as the ratio of total energy output for a certain period to the .... for the same period.

Ans: Total energy input

45. When the energy efficiency is calculated for a day of 24 hours, it is called .... efficiency.

Ans: All day

46. All day efficiency of a distribution transformer is .... its commercial efficiency.

Ans: Lower than

47. The transformers, used to step down the distribution voltage to a standard service voltage or from transmission voltage to distribution voltage are called as .... transformers.

Ans: Distribution

48. The transformers, which are generally used in generating stations or substations at each end of a power transmission line for stepping up or stepping down the voltage are known as .... transformers.

Ans: Power

49. An autotransformer is a .... winding transformer.

Ans: One

50. In an autotransformer, electrical power is transferred from primary to secondary partly by the process of transformation and partly by direct ………. connection.

Ans: Electrical

51. An autotransformer uses less winding material than a ………. transformer.

Ans: Two winding

52. An autotransformer has ………. efficiency than the equivalent two winding transformer.

Ans: Higher

53. Autotransformers with a number of tappings are used for starting induction and ………. motors.

Ans: Synchronous

54. In a transformer, two vertical bars are used to signify tight ………... coupling between the windings.

Ans: Magnetic

55. The winding, having higher number of turns, is called as ………. winding and the winding having lower number of turns is known as ………. winding.

Ans: High voltage, Low voltage

56. The transformer, which may receive energy at certain voltage and deliver it at the same voltage, is called as ………. transformer.

Ans: One to one

57. The ………. leakage flux is that flux, which links with the primary winding, but does not link with secondary winding.

Ans: Primary

58. An ideal transformer is basically an ………. transformer.

Ans: Imaginary

59. An ideal iron core transformer consists of two coils wound in the same direction on a ………. .

Ans: Common magnetic core

60. In an ideal transformer, magnetizing current (I_m) lags V_1 by the angle ....

Ans: **90°**

61. For an ideal transformer, input kVA is equal to .... kVA.

Ans: **Output**

62. The I_m is called as .... component, which magnetizes the ....

Ans: **Magnetizing; Core**

63. Magnetizing current sets up a flux in the core, so I_m is in phase with ....

Ans: **Φ_m**

64. I_w is called as active or .... component of no load current (I_0), which is in phase with applied voltage V_1.

Ans: **Wattful**

65. The construction of transformer is comparatively simple because there are no .... parts.

Ans: **Rotating**

66. In a transformer, the .... laminations are used as core material.

Ans: **Sheet**

67. Plain sheet steel tanks are generally used with small transformers of rating less than.......... kVA for normal voltages.

Ans: **50**

68. For small power transformers, it is preferable to use .... tanks.

Ans: **Radiator**

69. Cruciform shape is used in transformer core so as to reduce ....

Ans: **Winding copper**

70. The yoke cross-section is made larger compared to core cross section in transformer to reduce.......... current.

Ans: **Magnetizing**

71. The core of a high frequency transformer is an .... core.

Ans: **Air**

72. A transformer, which possesses highest value of reactance is called as .... transformer.

Ans: **Current**

73. An easily repairable transformer at site is of .... type.

Ans: **Core**

74. The area of surface of tank of a transformer can be increased by using .... tubes.

Ans: **Rounded section**

75. A transformer must not be connected to a .... source.

Ans: **DC**

76. In case of transformer, voltage per turn in secondary winding is.......... to the voltage per turn in primary winding.

Ans: **Equal**

77. The primary leakage flux is defined as that flux, which links with the primary, but not with ....

Ans: **Secondary**

78. Secondary leakage flux does not link with .... winding.

Ans: **Primary**

79. The ordinary efficiency is also called as .... efficiency.

Ans: **Commercial**

80. The all day efficiency of a transformer is the ratio of output in kWh to intake in ....

Ans: **kWh**

81. The efficiency is maximum at that load, which makes the copper loss equal to constant .... loss.

Ans: **Iron**

82. In a distribution transformer, core losses occur for the whole 24 hours, whereas copper losses occur only when the transformer is on ....

Ans: **Load**

83. The performance of a distribution transformer is more appropriately given by.......... efficiency.

Ans: All day or energy

84. .... efficiency is given as the ratio of the energy output to the energy input taken over a 24 hour period.

Ans: All day

85. Power transformers are designed to have maximum efficiency at or near.......... load.

Ans: Full

86. When transformers are used to measure voltage and current, these are known as ....

Ans: Instrument transformers

87. A single-phase autotransformer is a .... winding transformer, in which a part of winding is common to both high voltage and .... sides.

Ans: One; Low voltage

88. An autotransformer transfers electrical power between primary and secondary circuits partly through a magnetic link (induction) and partly ....

Ans: By direct electrical connection (conduction)

89. An autotransformer is smaller in size and cheaper compared to the.......... transformer of same output.

Ans: Two winding

90. The effective per unit impedance of an autotransformer is .... compared to a two winding transformer.

Ans: Smaller

91. Autotransformer may be used in boosting of supply voltage by a small amount in distribution systems to compensate ....

Ans: Voltage drop

92. Buchhloz relay has an oil tight container fitted with two internal floats, which operate .... switches connected to external alarm.

Ans: Mercury

93. The conservator is an .... drum supported on the transformer lid or on a neighbouring wall.

Ans: Air tight cylindrical

94. In oil forced natural cooling transformer, the heat developed in the transformer is passed to tank walls through .... where it is dissipated by natural circulation of air.

Ans: Oil

95. In a transformer, corrugated tanks have corrugations or fins welded between .... and the tank base.

Ans: Cover flange

96. In a single-phase transformer, major disadvantage of the transformer oil is its ....

Ans: Sludging

97. Animal and vegetable oils are not suitable for use in transformers because they form ....

Ans: Destructive fatty acids

98. Chlorinated diphenyl is a .... suitable for use in transformers.

Ans: Synthetic oil

99. In oil forced natural cooling, forced circulation of oil is done by a pump, while cooling is done by........

Ans: Natural circulation of air

100. The operation of Buchhloz relay depends on the fact that most internal faults within the transformer generate ....

Ans: Gases

101. Oil in a transformer serves the double purpose of .... and ....

Ans: Insulation; Cooling

102. In the construction of a transformer, sheet steel laminations are assembled to give a .... with a minimum of airgap included.

Ans: Continuous magnetic path

103. The thickness of laminations varies from 0.35 mm for a frequency of 50 Hz to .... mm for a frequency of 25 Hz.

Ans: **0.5**

104. Joints in the alternate layers in transformers are staggered so as to avoid the presence of ....

Ans: **Narrow gaps right through the cross section of the core**

105. In core type transformers, the windings surround a considerable part of .... whereas in shell type transformers, the core surrounds ....

Ans: **Core; A considerable portion of windings**

106. In core type transformers, both windings are located on the opposite legs of the core, but actually they are always interleaved to reduce ....

Ans: **Leakage flux**

107. In core and shell type transformers, individual laminations are cut in the form of long strips of .... shape.

Ans: **L, E and I**

108. To avoid .... at the joints, where laminations are butted against each other, the alternate layers are stacked differently.

Ans: **High reluctance**

109. In small size core type transformers, a simple rectangular core is used with cylindrical coils, which are either .... in form.

Ans: **Circular or rectangular**

110. The circular cylindrical coils are used in most of the core type transformers due to their ....

Ans: **Mechanical strength**

111. The cylindrical coils are wound in helical layers with various layers insulated from each other, by ....

Ans: **Paper, Cloth, Micarta board or cooling ducts**

112. Low voltage winding is placed near to the core because low voltage winding is easiest to ....

Ans: **Insulate**

113. In case of transformer, the reduction in core sectional area due to the presence of paper, surface oxide, etc. is of the order of ……….. approximately.

Ans: 10%

114. In transformer, stepping of core not only provides high space factor but also results in reduced ………. and consequent I^2R ………. .

Ans: Length of mean path; Loss

115. Cores and coils of transformers are provided with rigid mechanical bracing to prevent ………. .

Ans: Movement and possible insulation damage

116. In good quality transformers, good bracing reduces ………. .

Ans: Vibration and the objectionable humming sound

117. Cold rolled steel of high silicon content enables the designer to use considerably higher operating ………. with lower loss per kg.

Ans: Flux densities

118. Due to higher flux density, weight per kVA can be ………..

Ans: Reduced

119. In a good quality transformer, good transformer oil should be absolutely free from ………. .

Ans: Alkalies, Sulphur and moisture

120. The presence of even an extremely small % of moisture in oil is highly detrimental from the ………. viewpoint, because it reduces………. of oil.

Ans: Insulation; Dielectric strength

121. Sludging is caused due to exposure to………. during heating and results in formation of ………. that clogs the cooling ducts.

Ans: Oxygen; Large deposits of dark and heavy matter

122. In high voltage transformer installations, ………. or capacitor type bushings are employed.

Ans: Oil filled

123. Permeability of the core varies with the .... value of exciting current and the wave of exciting or magnetizing current is not truly..........

Ans: **Instantaneous; Sinusoidal**

124. When a transformer is loaded, then secondary voltage decreases due to ....

Ans: **Internal resistance and leakage reactance drops**

125. In transformer, secondary terminal voltage falls, as the load on the transformer is ...., when PF is lagging and it increases, when the PF is ....

Ans: **Increased; Leading**

126. A ferrite core has lower specific eddy current loss compared to an iron core because ferrite core has higher .... resistance.

Ans: **Electrical**

127. Maximum value of flux in an AC excited iron core is determined by ....

Ans: **Impressed voltage and frequency**

128. In distribution transformers, the core losses are less than .... losses.

Ans: **Full load cu**

129. Total amount of transformer loss does not depend on phase angle between .... i.e. it is independent of load power factor.

Ans: **Voltage and current**

130. At any volt-ampere load, the efficiency depends on power factor and will be maximum at .... power factor.

Ans: **Unity**

TRUE/FALSE

1. The transformer is a device, which transfers electrical energy from one electrical circuit to another electrical circuit.

Ans: **True**

2. Primary and secondary windings of a transformer are not connected electrically, but are coupled magnetically.

Ans: **True**

3. When energy transfer occurs at different voltage, the purpose of the transformer is merely to isolate two electrical circuits.

Ans: **False**

4. A step up transformer can not be used as a step down transformer.

Ans: **False**

5. A transformer, if connected across DC supply of the same voltage, is likely to be demaged.

Ans: **True**

6. The primary and secondary currents are directly proportional to their respective turns.

Ans: **False**

7. The voltage transformation ratio of a transformer is equal to inverse of its current transformation ratio.

Ans: **True**

8. In a transformer, primary winding consists of resistance and leakage reactance both.

Ans: **True**

9. In the primary and secondary windings of a transformer, leakage reactance is independent of load current.

Ans: **True**

10. Transformer has the highest possible efficiency out of all the electrical machines.

Ans: **True**

11. Transformer is the device, which may be used in low/ high voltage circuits.

Ans: **True**

12. The transformer has two general types named as core type and shell type.

Ans: **True**

13. Pulse transformers may be employed in digital computers.

Ans: True

14. In transformer, the coupling between primary and secondary winding is by magnetic field.

Ans: True

15. The magnetic core is basically a stack of thin silicon steel laminations.

Ans: True

16. Mutual flux is that flux, which links with secondary winding only.

Ans: False

17. A transformer consists of two windings, which are not insulated from each other and wound on a common core, made up of non-magnetic material.

Ans: False

18. The induced emf in the primary winding is nearly equal to the applied voltage V, but will oppose the applied voltage.

Ans: True

19. The emf (induced) in the windings will have the same frequency as that of the supply voltage.

Ans: True

20. When number of turns in secondary winding are more compared to primary winding turns, then it is called step down transformer.

Ans: False

21. In case of core type transformer, two iron paths are provided for the flux.

Ans: False

22. In core type transformers, sandwitched type coils are preferred.

Ans: False

23. In shell type transformers, two parallel magnetic paths are provided.

Ans: True

24. In transformers, the magnetic circuit is of laminated iron core.

Ans: True

25. In shell type transformer, LV and HV windings are wound over the central limb.

Ans: True

26. The vertical portions of the core are called as yoke.

Ans: False

27. Interleaved coils are employed in shell type transformers.

Ans: True

28. The lamination thickness must be high so as to prevent the eddy current losses.

Ans: False

29. CRGO (cold rolled grained oriented steel) reduces the magnetizing current taken by primary winding.

Ans: True

30. By using the cold rolling of laminated sheets, the grains are allowed to orient in the direction of rolling.

Ans: True

31. Hysteresis loss (in the core) occurs due to the fact that core is magnetized and demagnetized rapidly or alternatively.

Ans: True

32. An air core transformer may be used in radio devices.

Ans: True

33. The core may be made up of powdered ferromagnetic alloy in some devices.

Ans: True

34. In large size core type transformers, core cross-section must be circular for the economical use of core material.

Ans: True

35. Shell type transformers are generally employed for low voltage, low power levels.

Ans: True

36. Low power transformers are generally oil cooled.

Ans: False

37. Leakage flux is that flux, that leaks through core legs and non-magnetic material surrounding the core.

Ans: True

38. According to the principle of transformer, an emf is induced in a coil, whenever it links a changing flux.

Ans: True

39. In an ideal transformer, core losses have significant values.

Ans: False

40. In an ideal transformer, flux is confined to the magnetic core only.

Ans: True

41. The value of emf per turn in primary is greater compared to emf per turn in secondary winding.

Ans: False

42. In drawing the phasor diagram of transformer, at $t = 0$, flux is zero, so it is drawn in horizontal direction.

Ans: True

43. In the single-phase transformer, source voltage lags the flux by 90°.

Ans: False

44. Whenever load is connected to transformer in secondary side, then the compensating primary mmf must be greater compared to secondary mmf.

Ans: False

45. Ideally the flux in the core remains constant.

Ans: True

46. Primary volt-amperes must be equal to secondary volt-amperes, while dealing with transformers.

Ans: True

47. The value of instantaneous power input into primary must be greater as compared to instantaneous power output from the secondary.

Ans: False

48. In transformer, generated emf may also be called as counter emf or reaction emf.

Ans: True

49. In the calculation of transformer, impedance may be transferred to other side (winding) of the circuit, using voltage transformation ratio.

Ans: True

50. In an ideal transformer, power and volt-amperes remain unchanged (in both the sides of the core).

Ans: True

51. Magnetization curve of an actual transformer core is linear.

Ans: False

52. In transformer, the primary and secondary induced emfs (E_1 and E_2) are drawn in the same direction, because both of them are produced, due to the same flux.

Ans: True

53. The purpose of magnetizing current is to demagnetize the core.

Ans: False

54. The flux produced by the magnetizing current will be of sinusoidal shape because the supply voltage is sinusoidal.

Ans: True

55. Whenever, secondary winding is open circuited, the induced emf (E_2) will be same as terminal voltage (V_2).

Ans: True

56. The primary current drawn by primary winding at no load condition is nearly 3% to 5% of full load current.

Ans: True

57. Generally primary ampere turns of transformer remains greater compared to secondary ampere-turns.

Ans: False

58. Under no load condition of transformer, primary winding is open circuited.

Ans: False

59. I_0 is called as primary load component of current.

Ans: False

60. The portion of I_0 is used to supply iron losses (hysteresis and eddy current losses) in core and a small amount of I_0^2 Rt loss in the primary winding.

Ans: True

61. The phasor of self induced emf is drawn at 180° to phasor V_1.

Ans: True

62. The no load component of current may also be called as magnetizing current.

Ans: False

63. In a transformer, I_e is kept low by using laminated core and silicon steel.

Ans: True

64. Two emf (E_1 and E_2) leads the flux by 90°.

Ans: False

65. The value of secondary voltage will be different from secondary induced emf E, when voltage drop in winding is zero.

Ans: False

66. The phasor sum of I_2' and I_0 is equal to the primary winding current.

Ans: True

67. Leakage flux may induce some emf in two windings.

Ans: True

68. Since main flux and leakage flux both are present in case of a transformer, so it can be thought as a equivalent voltage drop due to two reactances.

Ans: True

69. Leakage flux are inversely proportional to the currents flowing in the respective windings.

Ans: False

70. The effect of leakage flux may be assumed as a equivalent to two reactors X_1 and X_2 connected in series with primary and secondary windings causing voltage drops.

Ans: True

71. By increasing the magnetic coupling between primary and secondary windings, leakage flux and leakage reactance may be reduced.

Ans: True

72. According to Lenz's law, self induced emf is equal and opposite to mutually induced emf.

Ans: False

73. Mutually induced emf and self induced emf are both induced by the same mutual flux.

Ans: True

74. I_m is called as wattless component of no load current.

Ans: True

75. The product of working component of no load current and supply voltage (V_1) is known as core loss of transformer.

Ans: True

76. In no load equivalent circuit of transformer, core loss is represented by R_0.

Ans: True

77. In case of practical transformer, the core has infinite permeability.

Ans: False

78. The current I_2' is known as secondary current referred to the primary winding.

Ans: True

79. In transformer, voltage regulation is expressed as the arithmetical difference in the secondary terminal voltage between no load and full rated load at a given PF with same value of applied primary voltage for rated as well as no load.

Ans: True

80. Circuit parameters are not essential for voltage regulation of any 1φ transformer.

Ans: False

81. The per unit resistance voltage drop may be called as per unit resistance R_e per unit.

Ans: True

82. In transformer, zero voltage regulation and maximum voltage regulation occur at leading and lagging PF respectively.

Ans: True

83. Eddy current loss is proportional to the square of load current.

Ans: False

84. Copper losses may vary as the square of the load current.

Ans: True

85. By using open circuit and short circuit test, efficiency as well as regulation of transformer may be obtained without loading the transformer.

Ans: True

86. By open circuit test of the transformer, cu loss may easily be obtained.

Ans: False

87. In performing short circuit test, LV side is generally short circuited and measurements are taken on the HV side.

Ans: True

88. Core loss is approximately proportional to the square of flux.

Ans: **True**

89. To perform a full load test on a transformer, case of maximum temperature rise must be kept in mind.

Ans: **True**

90. Efficiency of a transformer will be maximum at that particular condition, when core loss will be equal to cu loss.

Ans: **True**

91. Highest possible efficiency occurs at unity power factor.

Ans: **True**

92. All day efficiency of a transformer may also be called as energy efficiency.

Ans: **True**

93. An auto transformer possesses only one winding instead of two windings.

Ans: **True**

94. An autotransformer transfers electrical power between primary and secondary circuits partly by induction and partly by conduction.

Ans: **True**

95. In a transformer, power loss occurs due to power transformation.

Ans: **True**

96. An autotransformer possesses higher efficiency as compared to a two winding transformer.

Ans: **True**

97. The reduced internal impedance provides smaller fault current in autotransformer.

Ans: **False**

98. The ratio of the size of an autotransformer to the size of a two winding transformer is given by $(n-1)/n$, where n is the primary to secondary turn or voltage ratio.

Ans: **True**

99. There is no advantage of autotransformer compared to two winding transformer, when the transformation ratio is large.

Ans: True

100. Length of copper required and area of cross section of the winding are proportional to number of turns and current rating respectively.

Ans: True

101. In autotransformer, output voltage may be varied from zero to nearly 120% of the input voltage.

Ans: True

102. In the testing of transformers, three types of tests namely routine tests, type tests and some special tests, have been specified in IS: 2026 (part-I)-1977.

Ans: True

103. Measurement of noise level and power taken by fans and oil pumps, may come under the category of special test.

Ans: True

104. In case of winding of a transformer, insulation resistance must not be less than a certain minimum value at a particular temperature.

Ans: True

105. The saving in terms of weight of copper needed, if an autotransformer is employed, instead of a two winding transformer and voltage transformation ratio is 0.5, will be 50%.

Ans: True

VIVA VOCE QUESTIONS

Q. 1. What do you mean by magnetic field and magnetic reluctance?

Ans: Magnetic field is the effect of current in the surrounding space and magnetic reluctance is a type of resistance, which is offered by the magnetic circuit to mmf to create required flux in the circuit.

Q. 2. What do you mean by transformer? Explain it briefly?

Ans: A transformer is a static piece of apparatus by which electric power from one winding (one circuit) to other winding is transferred, through the medium of magnetic field without change in frequency.

Q. 3. What is the working principle of transformer?

Ans: It works on the principle of "Faraday's law of electro-magnetic induction". According to it, whenever a conductor links with the changing flux, an emf is induced in the conductor. This emf is also proportional to the rate of change of flux and the total number of conductors.

Q. 4. Why the frequency in a transformer does not change?

Ans: The flux in the cores has definite frequency and it will be same on the primary and secondary sides, so the question of changing the frequency does not arise.

Q. 5. When supply is given to low tension winding and load is connected to high tension winding, the transformer may be called as?

Ans: Step-up transformer

Q. 6. What is the main purpose of using core in a transformer?

Ans: The purpose is to reduce reluctance of the common magnetic circuit.

Q. 7. In a transformer, the leakage flux of each winding is proportional to the current in that winding, why?

Ans: The reason is that leakage paths do not saturate.

Q. 8. Why are the core bolts insulated in case of transformer?

Ans: The purpose is to prevent increase of eddy current loss.

Q. 9. Why are transformer stampings annealed before being used for the building of core?

Ans: It is done to reduce eddy current loss due to burning of edges.

Q. 10. In transformer, what is the physical basis, between two circuits linked by a common magnetic flux?

Ans: Mutual induction

Q. 11. In transformers, the core is made up of sheet steel laminations, why?

Ans: It is to provide a continuous magnetic path with a minimum of air gap included.

Q. 12. What are the various parts of the transformer?

Ans: These include:
Core, windings, transformer tank, conservator, breather, cooling arrangement, etc.

Q. 13. Which type of transformer, has the magnetic paths?

Ans: Shell type transformer

Q. 14. How the eddy current loss can be minimized in a transformer?

Ans: By laminating the core

Q. 15. Why the joints in the alternate layers are staggered?

Ans: It is to avoid the presence of narrow gaps.

Q. 16. Which factor determines the thickness of the stampings?

Ans: Frequency

Q. 17. On which factors, hysteresis loss depends?

Ans: It depends upon:
(a) Flux density   (b) Frequency
(c) Quality and amount of iron in the core

Q. 18. In transformer core, cruciform shape is preferred, why?

Ans: It is to reduce winding copper.

Q. 19. What is the main role of core in the transformer?

Ans: The main role is to provide an easy path for the magnetic lines of force and to reduce the eddy current loss.

Q. 20. Why the primary and secondary windings on the opposite lags are interleaved?

Ans: It is to reduce the leakage flux.

Q. 21. What are the various types of transformers?

Ans: The various types of transformers include:
(a) Core type transformer
(b) Shell type transformer
(c) Spiral core or wound core type transformer

Q. 22. While designing a transformer, the effective core area is reduced, why?

Ans: Due to laminations and insulation.

Q. 23. Why the low voltage (LV) winding is placed nearest to core?

Ans: Low voltage winding is easiest to insulate, so it is placed nearby the core.

Q. 24. Why the cores have stepped cross-section?

Ans: It provides high space factor and reduced length of the mean turn. That's why stepped core section is preferred.

Q. 25. What are the various advantages of using spiral core transformer?

Ans: The advantages include:
(a) Lower iron losses at higher flux densities
(b) Low manufacturing cost
(c) Lesser size per kVA rating
(d) Lesser weight
(e) More rigid construction, etc.

Q. 26. Why a solid homogeneous block is not used in a transformer?

Ans: Since this block can not reduce the production of eddy current, so eddy current losses occur in transformer. Thus instead of using simple block, laminated solid block is used.

Q. 27. What is the function of oil in transformer?

Ans: It provides cooling and winding insulation both.

Q. 28. What are the various factors, that affect the oil's dielectric strength?

Ans: These factors include:
(a) Dust   (b) Moisture
(c) Temperature, etc.

Q. 29. What is the function of breather in a transformer?

Ans: In large sized transformers, chamber called as breather is provided to permit the oil inside the tank to expand and contract as its temperature increases or decreases.

Q. 30. What do you mean by breather?

Ans: It is basically a pipe attached with the conservator. The air entering the transformer is made moisture free by letting it pass through a device, called as breather.

Q. 31. Why the conservators are employed in transformers?

Ans: To prevent the oil from having contact with air as well as moisture, conservators are employed.

Q. 32. In transformers, which fluid is used?

Ans: Now a days, instead of natural mineral oil, synthetic insulating fluid called as ASKARELS (trade name) are used.

Q. 33. In high voltage installations, which type of bushings are used?

Ans: Oil filled or capacitor type bushing.

Q. 34. What do you mean by limb in relation to transformer?

Ans: It is a vertical portion, where the windings are placed.

Q. 35. Which type of insulation is used between the laminations?

Ans: It may be insulating varnish or paper.

Q. 36. Why the conservator of transformer is not completely filled with oil?

Ans: It is not completely filled with oil because, due to increase in temperature, oil expands, which may create serious trouble.

Q. 37. What is the function of magnetizing current in transformer?

Ans: It is merely used to magnetize the core.

Q. 38. Why the corrugated sheets are provided in transformer?

Ans: They are provided to increase the heat dissipating area of tank.

Q. 39. Name the various cooling methods of transformer?

Ans: It includes:
(a) Air blast (AB) cooling
(b) Oil immersed forced air cooling
(c) Oil immersed forced water cooling
(d) Oil immersed natural air cooling
(e) Natural air cooling, etc.

Q. 40. What do you mean by voltage transformation ratio in case of transformer?

Ans: It is defined as the ratio of secondary voltage to primary voltage, i.e. E_2/E_1.

Q. 41. What is the relationship between voltage, current and turns in terms of transformation ratio?

Ans: $V_s/V_p = N_s/N_p = I_p/I_s$

Q. 42. What is the meaning of no load current of transformer?

Ans: When the primary voltage is applied to the transformer and secondary winding is kept open circuited, then primary draws, small amount of current called as no load current of transformer.

Q. 43. What do you mean by Buchhloz relay used in transformer?

Ans: Buchhloz relay is the relay, which is used for the safety purpose in transformer. It is placed in between main transformer tank and conservator.

Q. 44. What do you mean by emf equation of transformer?

Ans: It is an equation, shows relation between E, f, N and φ_m. In equation form, it can be expressed as:

$E = 4.44 f \varphi_m N$ volts

Q. 45. What do you mean by wattful and wattless components of no load current?

Ans: The component of I_0, i.e. $I_0 \cos\phi_0$ is used to overcome the hysteresis and eddy current losses, called as wattful component. Likewise wattless component is that component of I_0, which is used to create magnetic flux in the cores. It is also called as magnetizing component of no load current.

Q. 46. What is the relation between I_0, I_m and I_w?

Ans: $I_w = I_0 \cos\phi_0$ = working component of no load current
$I_m = I_0 \sin\phi_0$ = magnetizing component of no load current

Q. 47. If I_m and I_w are the two components of I_0, then what will be the relation between these components and I_0?

Ans: $I_0 = \sqrt{\left((I_m)^2 + (I_m)^2\right)}$

Q. 48. What is the relation between B_0, Y_0 and G_0, where B_0 = exciting susceptance, Y_0 = exciting admittance and G_0 = exciting conductance

Ans: $B_0 = \sqrt{\left((Y_0)^2 - (G_0)^2\right)}$

Q. 49. Why the magnetizing current is of importance for transformer?

Ans: The purpose of magnetizing current is to produce magnetic flux in the core. Actually for induction phenomenon, flux is the base and without induction, the secondary cannot induce any voltage, thus the transformer has no meaning without magnetizing current.

Q. 50. What are the two components of core loss?

Ans: It is made up of two components, called as
(a) Hysteresis loss (W_h)   (b) Eddy current loss (W_e)

Q. 51. What amount of primary rated voltage is used in short circuit test of a single-phase transformer?

Ans: A low voltage (nearly 5% to 10% of normal primary voltage) at correct frequency is used.

Q. 52. What are the other names of OC test and SC test?

Ans: The open circuit test is called as no load test and the short circuit test is known as impedance test.

Q. 53. Why the rating of transformer is given in kVA and not in kW?

Ans: Since we know that transformer has mainly two losses namely copper loss and iron loss. Total copper loss depends on current and iron loss on voltage. Hence, total transformer loss depends on volt-ampere (VA) and not on phase angle between voltage and current, i.e. it is independent of load power factor. That is why the rating of transformer is in kVA and not in kW.

Q. 54. What do we mean by regulation down and regulation up in a transformer?

Ans: The change in secondary terminal voltage from no load to full load is given as $_0V_2 - V_2$. When this change is divided by $_0V_2$, it is called as regulation "down" and if this is divided by V_2, it is called as regulation "up".

Q. 55. What is the general formula for finding regulation, in terms of percentage resistive and reactive drop?

Ans: % regulation = $V_r \cos\theta \pm V_x \sin\theta$ (approximately)

Q. 56. In a transformer, if applied primary voltage is raised from rated value V_1 to V_1', then what will be the regulation?

Ans: % regulation = $\dfrac{\left(V_1' - V_1\right)}{V_1} \times 100$

Q. 57. What are the formulas for obtaining the percentage resistance and the percent reactance at full load?

Ans: Actually percentage resistance at full load can be expressed as:

$\%R = (I_2 R_{02}/V_2) \times 100 = \%$ Cu loss at full load $= \%$ Cu loss and similarly percentage reactance at full load can be given as:

$\%X = (I_2 X_{02}/V_2) \times 100 = \%$ Reactive drop

Q. 58. If V_2, I_2, R_2, X_2 and PF is given, then how would you obtain % voltage regulation?

Ans: % Regulation = $\dfrac{\left(I_2 R_2 \cos\phi \pm I_2 X_2 \sin\phi\right)}{V_2} \times 100$

Q. 59. What is the disadvantage of Kapp regulation diagram?

Ans: Since the length of sides of the impedance triangle are very small as compared to the radius of circle, the diagram has to be drawn on a very large scale, if sufficiently accurate results are desired. It is the disadvantage of Kapp diagram.

Q. 60. For what purpose, Sumpner's back to back test may be used on 1φ transformer?

Ans: This test may be used for finding the efficiency, regulation and heating under loaded conditions, when performed on a single-phase transformer.

Q. 61. Why the core loss in a transformer remains nearly constant?

Ans: Since the core flux in a transformer remains practically constant for all the loads, so the core loss is practically the same at all loaded conditions.

Q. 62. What are the two components of core loss?

Ans: It has two components, namely hysteresis loss and eddy current loss.

Hysteresis loss = $W_h = \eta\,B_{max}^{1.6} f V$ watts and
Eddy current loss = $W_e = P\,B_{max}^{2} f^2 t^2$ watts

Q. 63. What is the necessary condition for obtaining maximum efficiency in a transformer?

Ans: The necessary condition is that iron loss and copper loss must be equal to each other. i.e. Cu loss = core (iron) loss

Q. 64. What is the formula for obtaining load kVA corresponding to max. efficiency?

Ans: Load kVA corresponding to maximum efficiency

$$= FL\,(kVA) \times \sqrt{(\text{Iron loss}/\text{FL copper loss})}$$

Q. 65. What types of losses occur in a transformer?

Ans: It includes:
(a) Iron (core) loss   (b) Copper loss
(c) Stray loss

Q. 66. Why iron losses are termed as constant losses?

Ans: Since the magnetic flux nearly remains constant throughout the working of transformer, so these losses are constant from no load to full load.

Q. 67. What do you mean by ordinary and all day efficiency of transformer?

Ans: The ordinary efficiency is the ratio of output power to input power (both in kW). Hence

$\eta_{ordinary}$ = power output/power input
Also all day efficiency is the ratio of output to input in a fixed time period and the total duration is taken 24 hours, so it may be expressed as:

All day efficiency = (output in kWh/input in kWh) × 100

Q. 68. By which test, core loss of a transformer may be measured?

Ans: Open circuit test or no load test.

Q. 69. Why the reading shown by wattmeter is considered as core loss in OC test?

Ans: The current flowing in the winding is very small in comparison to the full load current, but the cores of transformer are fully energized, i.e. having the normal working flux. That's why the reading shown by wattmeter is core loss.

Q. 70. Name those tests, which may be performed on a transformer so as to obtain efficiency?

Ans: It includes:
(a) Sumpner (Back to Back) test
(b) OC and SC tests

Q. 71. What is the special feature of Sumpner's test?

Ans: In this test, the transformer is loaded without consuming much power for several hours. Actually it is the only test (i.e. special feature) by which, copper loss, core loss, temperature rise and voltage regulation can be obtained.

Q. 72. Name the various tests, which must be performed before commencing the transformer?

Ans: These tests are:
Open circuit and short circuit test, phasing out test, polarity test, DC resistance test, insulation test, voltage ratio test, oil test, etc.

Q. 73. What do you mean by autotransformer?

Ans: It is that transformer, in which only one winding exists and part of this being common to both primary and secondary. In other sense, it can be said that a portion of the winding works as the primary and other as secondary.

Q. 74. Whether the value of current in primary and secondary is same or different in an autotransformer?

Ans: The current is obviously different in these windings and in a common portion, the current is the difference of primary and secondary currents.

Q. 75. For what purposes, autotransformer is used?

Ans: It may be used as:

(a) Interconnecting transformer in 132 kV/330 kV system.

(b) Auto starter transformer for starting the induction motors

(c) Furnace transformer, etc.

Q. 76. What is the saving in autotransformer in terms of wt. of cu in ordinary transformer?

Ans: Saving = $K \times$ (wt. of Cu in ordinary transformer), where K is a constant of autotransformer.

Q. 77. In autotransformer, how much power is transformed inductively?

Ans: $(1 - K) \times$ input

Q. 78. How much power is transferred conductively in an autotransformer?

Ans: $K \times$ input

Q. 79. What is the meaning of parallel operation of the transformer?

Ans: When the load increases in excess, then other transformer is connected in parallel with the existing one to share the load. It is called the parallel operation.

Q. 80. Which conditions are required to be satisfied for parallel operation of transformers?

Ans: These conditions include the following points:

(a) The transformers should be properly connected with regard to polarity.

(b) The voltage transformation ratio and % impedance drop must be same.

(c) The % impedance should be equal in magnitude and have same X/R ratio to avoid circulating currents and operation at different power factors.

(d) The equivalent impedances should be inversely proportional to the individual KVA rating of transformers.

Q. 81. What are the benefits of connecting transformers in parallel mode?

Ans: (a) Increasing demand of load can be met.

(b) Reliability, continuity of service and maximum operating efficiency may be obtained.

Q. 82. In a transformer, if the voltage applied is increased so as to keep V/F fixed, then how will the core loss and magnetizing current change?

Ans: In this case, core loss will increase but magnetizing current will remain same.

Q. 83. Why are the transformers placed in oil filled tanks?

Ans: Insulation materials commonly used for transformers are hygroscopic in nature, so the transformers can not be put in the open environment because, moisture will be absorbed by insulation and its insulating property will be lost. Thus to prevent ingress of moisture, the transformers are placed inside a sealed tank filled with sufficient quantity of oil to immerse the transformer leaving space on top through which it breathes via a moisture absorbing passage.

Q. 84. What do you mean by power and distribution transformers?

Ans: Those transformers, which are used in big firms and are used for any specific and particular purposes, are called as power transformers. These types of transformers are designed for maximum efficiency and less duration.

Similarly transformers, which are used to distribute electrical power to the consumers, are called as distribution transformers. These are designed to work for 24 hours.

Q. 85. For Sumpner (back to back) test on 1φ transformer, which type of transformers must be used?

Ans: In this test, two similar transformers must be used for carrying full load test.

6

Synchronous Generator (Alternator)

Q. 1. An alternator will run at the greatest possible speed, if it is wound for 50 Hz and .... poles:

(a) 6   (b) 2

(c) 4   (d) 8

Ans: (b)

Q. 2. The principle of operation of an alternator is:

(a) Mutual induction

(b) Conduction

(c) Electromagnetic induction

(d) Both (a) and (b)

Ans: (c)

Q. 3. Synchronous generator may also be known as:

(a) Synchronous condenser

(b) Synchronous compensator

(c) Alternator

(d) None of the above

Ans: (c)

Q. 4. Synchronous compensators are designed for ratings:

(a) Up to 100 KVAR

(b) Up to 1000 KVAR

(c) Up to 100 MVAR

(d) Up to 1000 MVAR

Ans: (c)

Q. 5. On the basis of construction, synchronous machines may be classified as:

(a) Rotating field type only

(b) Rotating armature type only

(c) Both (a) and (b)

(d) None of the above

Ans: (c)

Q. 6. Which of the following statements is true?

(a) In AC generators, armature is a revolving member

(b) Rotating armature type alternators are built in small ratings

(c) Synchronous machine may be classified as salient pole machine and cylindrical rotor machine

(d) All of the above

Ans: (d)

Q. 7. The benefits of revolving field structure include:

(a) Only two slip rings are needed for supply of direct current to the rotor

(b) In this design, exciting current is small

(c) Slip rings are of light construction

(d) All of the above

Ans: (d)

Q. 8. The purpose of stator frame in synchronous machine is to:

(a) Protect the armature windings

(b) Hold the armature stampings in position

(c) Circulate cold water for cooling in some cases

(d) Both (b) and (c)

(e) None of the above

Ans: (d)

Q. 9. The stator of synchronous generator gets overheated sometimes. It may be due to:

(a) Open circuit of phase

(b) Improper alignment of rotor

(c) Reduced supply current

(d) Both (b) and (c)

(e) Both (a) and (b)

Ans: (e)

Q. 10. In 3-phase synchronous machine, the slip rings are insulated for:

(a) Low voltage   (b) 1000 volts

(c) Very high voltage   (d) The range 500 V–2500 V

Ans: (a)

Q. 11. Three-phase alternators are star connected because:

(a) Higher value of terminal voltage is obtained

(b) Lower value of terminal voltage is obtained

(c) Losses are reduced to minimum

(d) Both (b) and (c)

Ans: (a)

Q. 12. Which of the following is a type of turbo alternator?

(a) Vertical type   (b) Horizontal type

(c) Inclined type   (d) Both (b) and (c)

(e) Both (a) and (b)

Ans: (e)

Q. 13. The armature core is built up of laminations of:

(a) Steel alloy   (b) Iron

(c) Special magnetic iron   (d) Both (a) and (b)

(e) Either (a) or (c)

Ans: (e)

Q. 14. Projecting pole type rotor of synchronous generator may be used in:

(a) Low and medium speed (engine driven) alternators

(b) Turbo generators

(c) Low and high speed (engine driven) alternators

(d) Medium speed alternators

Ans: (a)

Q. 15. **The cylindrical rotor construction of synchronous machine provides:**

(a) Quieter operation only

(b) More windage losses

(c) Better balance and less windage losses

(d) Quieter operation and more windage losses

Ans: **(c)**

Q. 16. **Smooth cylindrical type rotors are designed mostly for:**

(a) 6 poles

(b) 4 poles

(c) 2 poles or 4 poles

(d) 4 poles or 6 poles

Ans: **(c)**

Q. 17. **Which one of the following statements is not true?**

(a) The rotor produces the main field flux

(b) Stator carries the field winding

(c) 3-phase winding is put in the slots cut on the inner periphery of stator

(d) Stator is the stationary part of machine

Ans: **(b)**

Q. 18. **Salient pole rotors of synchronous machine have:**

(a) Distributed winding on the poles

(b) Concentrated winding on the poles

(c) Both lap and wave windings

(d) Wave winding

Ans: **(b)**

Q. 19. **A salient pole generator has:**

(a) Large diameter and short axial length

(b) Small diameter and short axial length

(c) Large diameter and large axial length

(d) Small diameter and large axial length

Ans: **(a)**

Q. 20. In a salient pole synchronous machine, the air gap is minimum and maximum at:

(a) Under the pole centres, in between the poles

(b) Above the pole centres, in between the poles

(c) Pole centres

(d) The centre of stator

Ans: **(a)**

Q. 21. A cylindrical rotor machine is also known as:

(a) Non-salient pole rotor machine

(b) Projecting pole rotor machine

(c) Salient pole machine

(d) Both (a) and (b)

Ans: **(a)**

Q. 22. The smooth rotor of a machine gives:

(a) Less windage losses   (b) Less noisy operation

(c) More windage losses   (d) Both (a) and (b)

(e) Both (b) and (c)

Ans: **(d)**

Q. 23. In a non-salient pole synchronous machine, the distribution of field mmf around the air gap is a:

(a) Flat topped stepped wave

(b) Traingular wave

(c) Square wave

(d) Cosine wave

Ans: **(a)**

Q. 24. In a synchronous machine (generating mode), the rotor field:

(a) Lags the stator field

(b) Leads the stator field

(c) Sometimes (a) and then (b)

(d) Nothing can be said perfectly

Ans: **(b)**

Q. 25. In a salient pole synchronous machine, the pole shoes are shaped in such a way that the air gap will have least value in the middle side and increases somewhat towards the edges. The construction:

(a) Reduces the harmonic content of the induced voltage wave

(b) Reduces 7th harmonic to zero

(c) Reduces 3rd harmonic to a minimum value

(d) None of these

Ans: (a)

Q. 26. The phasor addition of rotor and stator mmf is possible in a cylindrical rotor synchronous machine. It is so because:

(a) Both the mmfs are rotating in opposite directions

(b) Two mmfs are stationary w.r.t. each other

(c) Both the mmfs are rotating in same direction

(d) Nothing can be said about the matter

Ans: (b)

Q. 27. The stator core is laminated to reduce:

(a) Iron loss

(b) Hysteresis loss

(c) Weight of the machine

(d) Eddy current loss

Ans: (d)

Q. 28. When the synchronous generator is driven by steam turbine, it is called as:

(a) Steam generator   (b) Turbo generator

(c) Both (a) and (b)   (d) None of these

Ans: (b)

Q. 29. Salient pole field structure:

(a) Has adaptability to low and moderate speed operations

(b) Reduces windage losses

(c) Increases the speed of machine beyond N_s

(d) Reduces the speed below $N_s/7$

Ans: (a)

Q. 30. Non-salient type field structure provides:

(a) High value of speeds

(b) Quieter operations

(c) Both (a) and (b)

(d) Very low rotor speed

Ans: (c)

Q. 31. How many number of slots per pole per phase must exist for small and large machines?

(a) 3 to 4 for small machines and more than 5 for large machines

(b) 2 to 3 for small machines and 5 for large machines

(c) 5 to 10 for small machines and 15 for large machines

(d) 3 to 7 for small machines and 20 for large machines

Ans: (a)

Q. 32. Which of the following statements related to synchronous generator is incorrect?

(a) If the depth of the slot is increased, synchronous reactance in slot increases

(b) Total number of slots per pole per phase must be large to give sinusoidal generated emf

(c) In alternator, the field system is known as rotor

(d) The power rating of the exciter is 2% to 10% of power rating of synchronous generator

Ans: (d)

Q. 33. What must be the ratio of air gap to the pole pitch (approximate)?

(a) 0.08 to 0.025

(b) 0.008 to 0.02

(c) 0.2 to 0.5

(d) 0.8 to 1.0

Ans: (b)

Q. 34. When one slot per pole or slots equal to number of poles are used, then the winding is called as:

(a) Distributed winding   (b) Integral slot winding

(c) Concentrated winding (d) Double layer winding

Ans: **(c)**

Q. 35. When the conductors are kept in several slots under one pole, the winding is called as:

(a) Distributed winding   (b) Lap winding

(c) Spiral winding   (d) Hemitropic winding

Ans: **(a)**

Q. 36. In modern alternators, stators are wound for .... phase groups:

(a) 60°   (b) 30°

(c) 90°   (d) 310°

Ans: **(a)**

Q. 37. Coil pitch factor may also be known as:

(a) Coil span factor

(b) Pitch factor

(c) Ratio of phasor sum of coil side emfs/arithmetic sum of coil side emfs

(d) All of the above

Ans: **(d)**

Q. 38. Pitch factor is represented as:

(a) $\cos\alpha$ (b) $\cos\alpha/2$

(c) $2\cos\alpha/2$ (d) $\sin\alpha/2$

Ans: **(b)**

Q. 39. Which of the statements/expressions is wrong w.r.t. distribution factor?

(a) It may also be called as breadth factor

(b) $K_d = (\sin m\beta/2)/ (m\sin\beta/2)$, where m = slots/pole/phase, $m\beta$ = phase spread, β = angular displacement between the slots

(c) K_d = Arithmatic sum of component emfs/phasor sum of component emfs

(d) None of these

Ans: **(c)**

Q. 40. Which of the following expressions is correct?
Here K_f = form factor, K_p = pitch factor, K_d = distribution factor.

(a) E_{rms}/phase = $4 K_d K_p f \varnothing T$

(b) E_{rms}/phase = $4.44 f \varnothing T$

(c) E_{rms}/phase = $4.44 K_d K_p f \varnothing T$

(d) E_{rms}/phase = $4.44 K_p K_f \varnothing T$

Ans: **(c)**

Q. 41. In 3-phase synchronous generator, distributed armature winding:

(a) Increases rigidity

(b) Reduces rigidity

(c) Increases mechanical strength of the winding

(d) Both (a) and (c)

(e) Both (a) and (b)

Ans: **(d)**

Q. 42. In synchronous machines, which one of the following options will provide emf closer to sine waveform (nearly of sinusoidal form):

(a) Distributed winding in short pitch coils

(b) Concentrated winding in full pitch coils

(c) Wave winding

(d) Spiral winding

Ans: **(a)**

Q. 43. A coil having a span equal to 180° electrical is known as:

(a) Short pitch coil   (b) Full pitch coil

(c) Fraction pitch coil   (d) None of the above

Ans: **(b)**

Q. 44. The leakage fluxes may be categorized as:

(a) Tooth head leakage   (b) Overhang leakage

(c) Slot leakage   (d) All of the above

Ans: **(d)**

Q. 45. For cooling alternators of rating up to 50 MW:

(a) Circulating cold air system will be used

(b) Hydrogen cooling will be used

(c) Hollow water cooled conductors will be used

(d) None of the above

Ans: **(a)**

Q. 46. Which one of the following methods will be used for finding regulation of alternator?

(a) Ampere-turn or mmf method

(b) Potier method

(c) Synchronous impedance or emf method

(d) All of the above

Ans: **(d)**

Q. 47. Which one of the following tests is not performed on an alternator to obtain its performance?

(a) DC resistance test   (b) Open circuit test

(c) Short circuit test   (d) Voltmeter-ammeter test

Ans: **(d)**

Q. 48. The graph, which is plotted between armature current (I_a) and field current (I_f) is known as:

(a) Short circuit characteristic

(b) Open circuit characteristic

(c) Zero power factor characteristic

(d) None of the above

Ans: **(a)**

Q. 49. The characteristic curve, which is plotted between open circuit phase voltage and field current is called as:

(a) Open circuit characteristic

(b) Short circuit characteristic

(c) ZPF characteristic

(d) None of the above

Ans: **(a)**

Q. 50. Which one of the followings is not an assumption in the synchronous impedance method?

(a) Flux remains same under test and load conditions

(b) The magnetic reluctance to the armature flux is constant regardless of power factor

(c) The synchronous impedance is not constant

(d) All of the above

Ans: **(c)**

Q. 51. Which of the following methods is known as ampere-turns method?

(a) Magnetomotive force (mmf) method

(b) Synchronous impedance method

(c) Potier method

(d) None of the above

Ans: **(a)**

Q. 52. The zero power factor characteristics can be obtained by loading the alternator using:

(a) Induction motor

(b) Commutator motor

(c) Lamp load

(d) Resistive and capacitive load

Ans: **(c)**

Q. 53. The following data are needed for potier method of determination of voltage regulation of an alternator:

(a) Zero PF curve

(b) Short circuit data

(c) No load curve and zero PF curve

(d) No load curve and short circuit data

Ans: **(c)**

Q. 54. The regulation of a synchronous generator may be defined as:

(a) $(V - E_0)/E_0 \times 100$

(b) $(E_0 - V)/E_0 \times 100$

(c) $V/E_0 \times 100$

(d) $(E_0 - V)/V \times 100$

Ans: (d)

Q. 55. Electromotive force method provides more regulation compared to ampere-turns method. It is so because in this case:

(a) Saturation effect is ignored

(b) Saturation effect is considered

(c) Armature reaction is considered zero

(d) None of the above

Ans: (a)

Q. 56. The voltage regulation of a synchronous generator will be negative, when:

(a) The machine is running at high loads

(b) The machine is running at low loads

(c) The load power factor is leading

(d) The load power factor is zero

Ans: (c)

Q. 57. An alternator is supplying power to infinite bus bars at unity power factor. When its excitation is increased, it will supply:

(a) The same power at a lagging power factor

(b) More power at zero power factor

(c) More power at leading power factor

(d) None of the above

Ans: (a)

Q. 58. An alternator is operating at constant load condition and its excitation is adjusted to provide unity PF current. When the excitation is increased, the power factor will be:

(a) Of lagging nature

(b) Of leading nature

(c) Zero

(d) None of the above

Ans: **(a)**

Q. 59. A 3-phase alternator having constant steam input gives power to an infinite bus at a lagging power factor. When the excitation is increased:

(a) Power angle decreases

(b) Power angle increases

(c) Power factor and power angle decreases

(d) Power factor decreases

(e) Power factor increases

Ans: **(c)**

Q. 60. In a salient pole synchronous machine, where d-direct axis, q-quadrature axis, E_f – excitation emf, then:

(a) I_q is in the phase with E_f and I_d at 90° to E_f

(b) I_q and I_d are at 90° to E_f

(c) I_d is in the phase with E_f

(d) I_q is at 90° to E_f

Ans: **(a)**

Q. 61. In a salient pole synchronous machine:

(a) $x_q > x_d$

(b) $x_q < x_d$

(c) $x_d = 0$

(d) $x_q = 0$

Ans: **(b)**

Q. 62. Which one of the followings is the reason for parallel operation of alternator?

(a) To meet the increasing future load demand

(b) Large number of alternators can supply greater amount of load

(c) The cost of energy generated is reduced

(d) All of the above

Ans: **(d)**

Q. 63. Which one of the following statements is true?

    (a) Synchronising lamp method is very costly

    (b) In the dark lamp process of synchronization, the lamp filaments might burn out

    (c) The frequency of the incoming machine must not be equal to frequency of live bus-bar

    (d) None of the above

Ans: (b)

Q. 64. The load sharing between two steam-driven alternators (in parallel operation) can be adjusted:

    (a) By varying steam supply to their prime movers

    (b) By changing the speed of machines

    (c) By changing field current of alternators

    (d) By varying power factor of any one alternator

Ans: (a)

Q. 65. The purpose of damper winding is:

    (a) To eliminate harmonic effects

    (b) To provide a low resistance path for the currents due to unbalancing of voltage

    (c) To reduce the starting current only

    (d) Both (a) and (b)

Ans: (d)

MISCELLANEOUS

Q. 1. In a synchronous machine, the torque angle (operating from a constant voltage bus) is the space angle between:

    (a) Rotor mmf wave and stator mmf wave

    (b) Flux density and stator mmf wave

    (c) Current and emf generated

    (d) Emf and rotor mmf wave

Ans: (a)

Q. 2. In the slip test of a synchronous generator, what will be the voltage across the open field circuit terminals?

(a) AC voltage of slip frequency

(b) AC voltage of twice the supply frequency

(c) AC voltage of normal frequency

(d) None of the above

Ans: (a)

Q. 3. In the synchronization process, an infinite bus bar must maintain:

(a) Constant frequency   (b) Constant voltage

(c) Infinite frequency   (d) Both (b) and (c)

(e) Both (a) and (b)

Ans: (e)

Q. 4. What must be the pitch of the coils so as to eliminate rth harmonic from the induced emf in a phase of a synchronous machine?

(a) $2r/(r-1)$th fraction of full pitch

(b) $(r-1)/r$th fraction of short pitch (¾th full pitch)

(c) $(r-1)/r$th fraction of full pitch

(d) None of the above

Ans: (c)

Q. 5. What must be the voltage regulation of a synchronous generator, if the data is as follows:

Power factor = 0.75 leading

No load induced emf = 2400 V

Rated terminal voltage = 3000 V

(a) –20%   (b) –30%

(c) 27.7%   (d) –27.7%

Ans: (a)

Q. 6. Two reactance theory of synchronous machines was proposed by:

(a) Andre blondel   (b) James brown

(c) Lenz   (d) Kirchoff

Ans: (a)

Q. 7. In case of synchronous machines, stray load losses are due to:

(a) Skin effect in armature conductors

(b) Armature leakage flux as eddy current loss in core and plates

(c) Eddy currents in the armature conductors

(d) All of the above

Ans: **(d)**

Q. 8. What must be the value of reactive power output in a round rotor alternator under the conditions of maximum active power output?

(a) $-3E_f^2/X_s$ (b) $-3V_f^2/X_s$

(c) $3X_s/V_f^2$ (d) $-3E_f^2/2X_s$

Ans: **(b)**

Q. 9. In cylindrical and salient pole synchronous machines, the maximum power will be achieved at load angles of:

(a) $90°, < 90°$ (b) $180°, 35°$

(c) $180°, < 90°$ (d) $35°, < 180°$

Ans: **(a)**

Q. 10. A 3-phase constant voltage, constant frequency bus is fed power from a turbo-alternator set. What will happen if the steam supply to the set is cut off?

(a) The set will stop

(b) The set will continue to run at rated speed in the same direction

(c) The set will run at ½ N_s

(d) None of the above

Ans: **(b)**

Q. 11. In a 3-phase synchronous generator, various methods of voltage regulation are

(1) Synchronous impedance method

(2) Saturated synchronous reactance method

(3) New ASA method

(4) Ampere-turns method

What must be the correct sequence of the ascending order of the values of the regulation obtained by these methods?

(a) 1, 2, 3, 4   (b) 2, 3, 4, 1

(c) 4, 3, 2, 1   (d) 4, 2, 3, 1

Ans: (c)

Q. 12. Maximum power output will be given by a cylindrical rotor synchronous generator, when:

(a) Load angle = $45°$

(b) Power factor is nearly zero

(c) Load angle = synchronous impedance angle

(d) Power factor is nearly 0.5

Ans: (c)

Q. 13. A 2 MVA, 3-phase, 8 pole synchronous generator is connected to 6000 V, 50 Hz bus bars and has a synchronous reactance of 4 ohms per phase. What will be the synchronising power?

(a) 628318 W/mech degree

(b) 528318 W/mech degree

(c) 52000 W/degree

(d) 25025 W/degree

Ans: (a)

Q. 14. The data of a turboalternator is given as follows:

$P = 2, f = 50$ Hz, Total phases = 3, capacity = 100 MVA, kV rating = 33 kV, Moment of inertia = 10^6 kgm^2, synchronous reactance = 0.5 pu.

In this case, turboalternator is connected to the infinite bus. What will be the natural frequency of oscillation?

(a) 7.75 s^{-1} (b) 7.874 s^{-1}

(c) 8.52 s^{-1} (d) 7.93 s^{-1}

Ans: (b)

Q. 15. Which one of the following statements is wrong?

(a) Short circuit ratio = Field current for rated open circuit voltage/Field current for rated short circuit current

(b) $E_f = 4.44 \, K_w f \varnothing T_{ph}$

(c) $E_f \propto$ Reluctance of air gap/field mmf per pole

(d) SCR $\propto$ Air gap reluctance

Ans: (c)

Q. 16. In a two layer winding, the stator slots will be equal to the number of:

(a) Coils   (b) Wires

(c) Total conductors   (b) Total turns

Ans: (a)

Q. 17. In a synchronous generator, harmonics (developed in the emf wave) may be reduced by:

(a) Using distributed winding

(b) Chamfering the salient pole tips

(c) Either (a) or (b)

(d) Both (a) and (b)

Ans: (d)

Q. 18. In a synchronous generator, terminal voltage reduces due to armature reaction. It is actually encountered by:

(a) Automatic voltage regulator

(b) Wave winding connection

(c) Auxiliary winding

(d) Damper winding

Ans: (a)

Q. 19. What will happen if the windings of alternator gets overheated?

(a) The life of machine reduces

(b) Value of power factor increases

(c) Starting current increases and then decreases

(d) Nothing can be said without knowing rating

Ans: (a)

Q. 20. What will be the values of direct axis and quadrature axis synchronous reactances of an alternator (by using slip test)?

(a) $X_d = V_{max}/I_{min}; X_q = I_{min}/V_{max}$
(b) $X_d = V_{max}/I_{min}; X_q = V_{min}/I_{max}$
(c) $X_d = 2I_{min}/V_{max}; X_q = V_{min}/I_{min}$
(d) $X_d = I_{min}/2V_{max}; X_q = V_{min}/2I_{max}$

Ans: **(b)**

Q. 21. In an alternator, armature and field windings are provided on stator and rotor respectively. In a particular situation, when machine is running under steady state condition, the air gap field:

(a) Is rotating at synchronous speed
(b) Is rotating at ½ (synchronous speed)
(c) Is rotating at synchronous speed in the direction of rotor rotation
(d) Is rotating at synchronous speed in the opposite direction of rotor rotation

Ans: **(c)**

Q. 22. In a physical system, electromagnetic torque developed tends to:

(a) Reduce impedance
(b) Increase reluctance
(c) Increase field energy and co-energy at constant current
(d) Increase field energy and reduce co-energy

Ans: **(c)**

Q. 23. In an alternator, the armature flux, at unity power factor (load) will be:

(a) Cross magnetising
(b) Demagnetising
(c) In phase with voltage
(d) In phase with current and voltage both

Ans: **(a)**

Q. 24. What must be the value of space angle between two consecutive phases in deg. electrical and deg. mechanical in a 3-phase, 6 pole winding?

(a) 100° and 40° respectively

(b) 120° and 60° respectively

(c) 120° and 75° respectively

(d) 120° and 40° respectively

Ans: (d)

Q. 25. What must be the pitch factor for 5/6 short pitch coil?

(a) 0.866

(b) 0.923

(c) 0.577

(d) 0.966

Ans: (d)

Q. 26. Which one of the followings is the cause of overheating of electrical machine?

(a) Excessive load

(b) Single phasing

(c) Poor ventilation facility

(d) All of the above

Ans: (d)

Q. 27. When the excitation of an alternator is changed, what will happen to voltage of an alternator connected to infinite bus bars?

(a) The voltage increases

(b) First the voltage increases up to a certain point and then decreases

(c) The voltage remains same

(d) The voltage decreases

Ans: (c)

Q. 28. Two alternators are operating in parallel mode. They are supplying a load of 8 MW at 0.8 power factor lagging. The power output of first machine is adjusted to 5000 kW by changing its steam supply and its power factor is varied to 0.9 lagging by changing its excitation. What must be the value of power factor of second alternator?

(a) 0.52   (b) 0.64

(c) 0.80   (d) 0.72

Ans: **(b)**

FILL IN THE BLANKS

1. Alternator may also be known as ....

Ans: **Synchronous generator**

2. In alternator, the armature windings are placed on the .... slots.

Ans: **Stator**

3. Diesel engines may be used as prime-movers for .... rating .... generators.

Ans: **Low; Synchronous**

4. Synchronous compensator may also be known as ....

Ans: **Phase modifier**

5. Now-a-days, all medium and large machines are always constructed with .... field.

Ans: **Revolving**

6. In an alternator, large sized generators for highest speeds are of the .... rotor type.

Ans: **Cylindrical**

7. The maximum speed will be .... rpm for a turbo-generator having 2 poles and 50 Hz frequency.

Ans: **3000**

8. In synchronous machines, semi closed slots do not permit the use of .... coils.

Ans: **Form wound**

9. The exciter armature may be connected directly to the alternator .... winding.

Ans: **Field**

10. In an exciter, its voltage rating lies between .... volts.

Ans: **125 and 250**

11. In large machines, the field windings are made of rectangular .... wound on edge.

Ans: **Copper strip**

12. In salient pole type field system, nearly .... of pole pitch is covered by pole shoes.

Ans: **2/3rd**

13. Smooth cylindrical type rotor provides .... balance and .... operation.

Ans: **Better; Quieter**

14. Damper windings may prevent .... phenomenon in case of synchronous generators.

Ans: **Hunting**

15. .... layer winding is referred to as chain or .... winding.

Ans: **Single; Concentric**

16. Chain windings may be used for large .... voltage systems.

Ans: **High**

17. .... pitched coils save copper of .... connections.

Ans: **Short; End**

18. Distribution factor is the ratio of .... sum of coil emfs to arithmetic sum of coil emfs.

Ans: **Vector**

19. For 3rd harmonic component, pitch factor (K_c) will be ....

Ans: **$\cos 3\alpha/2$**

20. The value of pitch factor for an alternator having 9 slots per pole and coil span equal to 8 slot pitches will be ....

Ans: **0.985**

21. For 1000 MW range machines, hollow .... are required for cooling purpose.

Ans: **Water cooled conductors**

22. The leakage flux is normally independent of ....

Ans: **Saturation**

23. In synchronous generators, load power factor has a great effect on ....

Ans: **Armature reaction**

24. The overall change in the magnitude of terminal voltage of an alternator depends not only on the load but on .... also.

Ans: **Load power factor**

25. Two methods of voltage regulation, MMF method and Potier method are also known as .... and .... respectively.

Ans: **Ampere-turns method; Zero power factor method**

26. The open circuit characteristic of synchronous generator is just like .... curve.

Ans: **B–H**

27. The open circuit characteristic of an alternator is plotted between field current and ....

Ans: **Open circuit voltage**

28. In ampere-turns method, the demagnetising armature AT on full load are equal and .... to the field ampere-turns needed to create .... on short circuit.

Ans: **Opposite; Full load current**

29. In cylindrical rotor machines, Potier reactance is nearly same as that of .... reactance.

Ans: **Armature leakage**

30. The zero power factor (lagging) characteristic may be plotted between ....

Ans: **Terminal voltage and excitation current**

31. Salient pole machines have two axes, namely .... and .... axes.

Ans: **Direct; Quadrature**

32. The axis passing through the centres of the interpolar space is called as .... axis.

Ans: **Quadrature**

33. In salient pole synchronous machines X_q is nearly equal to .... times ....

Ans: **0.6 to 0.7; X_d**

34. The direct axis reactance and quadrature axis reactance are .... and .... respectively.

Ans: $X_d = V_{max}/I_{min}$; $X_q = V_{min}/I_{max}$

35. When a supply system with large number of alternators is connected in parallel, it is referred to as .... bar.

Ans: **Infinite bus**

36. For satisfactory parallel operation of alternators, frequency of the generated voltage of the incoming alternator must be .... to the bus bar frequency.

Ans: **Equal**

37. Synchronization of alternators can be achieved by using a synchroscope or .... lamp system.

Ans: **Three (One dark and two bright)**

38. After synchronization process, a synchronous machine just .... on the bus bar.

Ans: **Floats**

39. Each synchronous machine has a .... time period of free oscillation.

Ans: **Natural**

40. Damper windings may also be known as .... windings.

Ans: **Amortisseur**

41. .... may also be placed on the engine governors so as to reduce tendency to hunt.

Ans: **Dash-pots**

42. .... efficiency and reliability, etc.may be obtained by the .... operation of alternators.

Ans: **Higher; Parallel**

43. Hunting may also be reduced by employing .... flywheels.

Ans: **Heavy**

44. The active power loading of an alternator (operating an infinite bus) may be controlled by controlling the .... to it.

Ans: Input power

45. The speed-load curve of the prime-mover of synchronous generator must be a .... curve.

Ans: Drooping

46. .... is an instrument, which indicates the difference of phase and .... between two voltages.

Ans: Synchronoscope; Frequency

47. Alternators are built in sizes .... compared to DC generators.

Ans: Much larger

48. When the excitation of an alternator, operating in parallel with other alternators is increased above normal value of excitation, its power factor changes in the .... direction.

Ans: Lagging

49. The overall process of checking the phase sequence and getting it correct is called as ....

Ans: Phasing out of alternator

50. In synchronous generators, the rotor field windings are excited by .... supply fed through ....

Ans: DC; Slip rings

51. When the load on alternator is inductive, terminal voltage will .... with increase in load.

Ans: Reduce

52. Magnetisation characteristic of a synchronous generator may also be known as .... characteristic.

Ans: No load

53. Hydrogen cooling may be used for alternators, as hydrogen avoids the .... of the insulating material.

Ans: Oxidation

54. In alternators, the power factor is dependent upon ....

Ans: Load

TRUE / FALSE

1. Alternators operate on the principle of electromagnetic induction.

Ans: True

2. In synchronous generators, the magnetic poles are excited from AC current.

Ans: False

3. In AC machines, stator frame is used for holding the armature stampings.

Ans: True

4. The stator core is laminated to reduce hysteresis losses.

Ans: False

5. Turbo generators are characterised by large diameter and short axial length.

Ans: False

6. The construction of a synchronous generator does not depend upon the type of primemover used to rotate the rotor.

Ans: False

7. In synchronous machines, rotor carries the field poles.

Ans: True

8. The rating of AC machinery is determined by their heating and hence by losses in them.

Ans: True

9. Pilot exciter and the main exciter are driven by the synchronous machine main shaft.

Ans: True

10. In a 3-phase alternator, the phase sequence can be reversed by interchanging the terminals of its field winding.

Ans: False

11. Rotating armature type alternator is economical for high voltage generators.

Ans: False

12. In rotating armature type alternator, slip rings and the brush gears are of heavy construction.

Ans: **False**

13. Hydrogenerators are driven by steam turbines.

Ans: **False**

14. Water wheel generators may also be called as hydrogenerators.

Ans: **True**

15. The thickness of stator core laminations is usually 0.9 mm.

Ans: **False**

16. The open slots facilitate in removal and replacement of defective coils.

Ans: **True**

17. Brushless excitation system means the combination of DC exciter and rectifiers.

Ans: **False**

18. The salient pole type field structure is used almost for high speed alternators.

Ans: **False**

19. In salient pole arrangement, the pole faces are so shaped that the radial air gap length increases from the pole tips to pole centre.

Ans: **False**

20. In the rotor of turbogenerators, chromium nickel steel may be used.

Ans: **True**

21. The non-salient field structure has robust construction and high operating speed.

Ans: **True**

22. When the conductors are placed in various slots, the concentrated winding is obtained.

Ans: **False**

23. Concentric windings may be used in slow speed large diameter alternators.

Ans: **True**

24. When the coil span of the winding is less than 180 electrical space degrees, the winding is known as fractional pitch winding.

Ans: **True**

25. With double layer arrangement, fractional slot winding is practicable.

Ans: **True**

26. Coil pitch factor may be represented by K_d.

Ans: **False**

27. In cylindrical rotor machines, air gap length is uniform throughout.

Ans: **True**

28. Higher order harmonics can be reduced by using integral slot winding.

Ans: **False**

29. Alternator's prime-mover has a rating independent of power factor.

Ans: **True**

30. For any value of power factor of load, the armature reaction has cross magnetising component proportion to $I \cos\phi$.

Ans: **True**

31. The synchronous reactance is basically the addition of fictitious reactance (X_a) and leakage reactance (X_L).

Ans: **True**

32. Load characteristic is drawn between terminal voltage and field current of an alternator.

Ans: **False**

33. By using short circuit test of an alternator, the synchronous impedance can be obtained.

Ans: **False**

34. In the open circuit test of alternator, the armature winding circuit is kept open and the machine is run at rated speed.

Ans: **True**

35. In the open circuit test, rheostat is inserted in the DC field circuit to keep the current low.

Ans: **False**

36. Short circuit ratio of synchronous machine is inversely proportional to synchronous reactance.

Ans: **True**

37. For low speed salient pole generators, short circuit ratio is in the range of 0.5 and 0.60.

Ans: **False**

38. The voltage regulation (pu) of an alternator is the ratio of the difference of no load terminal voltage to full load terminal voltage.

Ans: **True**

39. Synchronous impedance method may also be called as pessimistic method.

Ans: **True**

40. Ampere-turns method for finding synchronous impedance of an alternator is also known as optimistic method.

Ans: **True**

41. The zero power factor (lagging) characteristic is the curve, which is drawn between terminal voltage and excitation current.

Ans: **True**

42. In a salient pole machine, the reluctance of magnetic paths on which the emf acts, are same along the direct and quadrature axes.

Ans: **False**

43. In the construction of two reactance diagram of synchronous machine, the angle between E_0 and V is known as load or power angle.

Ans: **True**

44. By using the concept of constant flux linkage, the effects of short circuit currents can be determined.

Ans: True

45. Core loss in an alternator includes ventilation loss and stray power loss.

Ans: False

46. There exists no ventilation problem in turbo generators.

Ans: False

47. The life of AC machines is the life of the insulation of windings.

Ans: True

48. All types of losses occur in a synchronous generator except field losses.

Ans: False

VIVA VOCE QUESTIONS

Q. 1. Which parts/elements are needed in a synchronous generator?

Ans: The elements in synchronous generator are as follows:
(1) Stator and rotor   (2) Prime mover
(3) Exciter

Q. 2. Which type of rotor construction is available for synchronous generators?

Ans: Synchronous generator has two types of rotor construction:
(a) Salient pole rotor and   (b) Cylindrical rotor

Q. 3. Which type of alternator is more stable?

Ans: Salient pole type alternator.

Q. 4. In which mode, the synchronous machines may be classified on the basis of construction?

Ans: It may be classified as
(a) Rotating armature type and
(b) Rotating field type

Q. 5. On which factor, the construction of an alternator depends?

Ans: The construction of an alternator depends upon the type of prime mover, used to rotate the rotor.

Q. 6. For which type of alternators, salient pole type field structure is used?

Ans: Slow and moderate speed alternators.

Q. 7. The rotating part of turbo generator is subjected to high mechanical stresses, why ?

Ans: It is due to high peripheral speed.

Q. 8. What are the advantages of rotating field type alternator over rotating armature type alternator?

Ans: These advantages are as follows:
  (a) Since the armature windings are stationary, they are not subjected to vibration
  (b) Rotating field is comparatively light and can be constructed for high speed rotation
  (c) As the exciting current is relatively small, the slip rings and the brush gear need be of light and simple construction
  (d) It becomes easier to insulate stationary armature winding for high voltages because large space is available on the stator portion for providing more insulation as the stator is outside the rotor, etc.

Q. 9. On the basis of prime movers used, what are the various types of AC generators?

Ans: It includes:
  (a) Turbo generators   (b) Hydrogenerators
  (c) Engine driven generators

Q. 10. What do we mean by hydrogenerators?

Ans: Actually salient pole alternators driven by water turbines are called hydrogenerators. They may also be known as hydro-alternators.

Q. 11. Which machine is called as non-salient pole rotor machine?

Ans: Cylindrical rotor machine.

Q. 12. What do you mean by exciter?

Ans: In synchronous machines, direct current is needed to excite the field winding provided on the rotor. It is actually provided to the rotor field by a DC generator called exciter.

Q. 13. What are the benefits of using brushless excitation system?

Ans: This system needs less maintenance because brushes and slip rings are absent. Also the amount of power loss is reduced.

Q. 14. What do we mean by distributed winding?

Ans: When the conductors are placed in several slots under one pole, the winding is known as distributed winding.

Q. 15. What are the benefits of distributed winding?

Ans: The benefits are as follows:
   (a) Since a number of small slots are evenly spaced, so the core is better utilized
   (b) Some distorting harmonics can be eliminated
   (c) The harmonic emfs are reduced and so the overall waveform is improved, etc.

Q. 16. What do you mean by balanced winding?

Ans: When the total number of coils per coil group is a whole number, the winding is called as balanced winding.

Q. 17. What is the meaning of coil group?

Ans: Coil group is the product of total number of phases and number of poles.

Q. 18. What do you mean by integral and fractional slot winding?

Ans: When the number of slots per pole per phase is an integer, the winding is called as integral slot winding. Also when the number of slots per pole per phase is a fractional number, the winding is known as fractional slot winding.

Q. 19. What is coil span factor?

Ans: Coil span factor is also known as coil-pitch factor or pitch factor. It is represented as K_c or K_p. It is the ratio of phasor sum of induced emfs per coil to the arithmetic sum of induced emfs per coil.

Q. 20. What is the breadth factor?

Ans: It is also known as distribution factor and represented by K_d. It is the ratio of phasor sum of coil voltages per phase to arithmetic sum of coil voltages per phase.

Q. 21. What is emf equation of alternator?

Ans: It is given as $E_{rms/phase} = 4.44 f K_d K_p \varphi T$ volts

Q. 22. What are the methods for elimination of harmonics from the output voltage waveform of AC machine ?

Ans: The methods are:
(a) By using fractional-slot winding
(b) By skewing the poles through one slot pitch
(c) By keeping larger air-gap length
(d) By short chording the windings of armature

Q. 23. Which factor governs the rating of electrical machine?

Ans: It is governed by the temperature rise due to the internal losses of the machine.

Q. 24. What is synchronous reactance?

Ans: It is the sum of two reactances in a synchronous machine. First one is leakage reactance (X_L) and other is fictitious reactance (X_a).

Q. 25. Which reasons are responsible for the change in terminal voltage of an alternator at load?

Ans: The reasons are the voltage drops due to:
(a) Effective armature resistance

(b) Armature leakage reactance

(c) Armature reaction

Q. 26. By which tests, the synchronous impedance can be obtained?

Ans: These tests are:

(a) Open circuit test   (b) Short circuit test

Q. 27. What is SCR in synchronous machines?

Ans: The short circuit ratio is abbreviated as SCR. It is the ratio of the field current to produce rated voltage on open circuit to field current needed to circulate rated current on short circuit, when the machine is running at synchronous speed.

Q. 28. What do you mean by the regulation of synchronous generator?

Ans: The variation of terminal voltage from no load to full load expressed per unit or % of full load voltage is known as regulation of synchronous generator.

Q. 29. Name the various methods for finding voltage regulation of an alternator?

Ans: The methods for finding voltage regulation are:

(a) Synchronous impedance (EMF) method

(b) Ampere-turns method (MMF)

(c) Potier method

Q. 30. What is zero power factor full load voltage characteristic?

Ans: It is actually a curve drawn between terminal voltage and excitation, while the machine is to run on synchronous speed and providing full load current at zero power factor.

Q. 31. What is the formula for determining voltage regulation?

Ans: It is expressed as follows:

% Regulation = $(E_0 - V)/V \times 100$

Q. 32. The short circuit characteristic of a synchronous generator is linear, why?

Ans: This characteristic is basically a straight line passing through the origin because the net excitation is so small that there is no saturation in the magnetic circuit.

Q. 33. Which method gives the pessimistic value of the regulation of a synchronous generator?

Ans: Synchronous impedance method (EMF method).

Q. 34. What is the use of Potier triangle?

Ans: It is a triangle (made in ZPFC), which is used for finding voltage regulation of an alternator.

Q. 35. What is an air gap line?

Ans: Air gap line is the line which is tangent to open circuit characteristic (OCC).

Q. 36. Why the reactance of salient pole synchronous machine varies with the rotor position ?

Ans: It is because the salient pole machine has non-uniform air gap.

Q. 37. What do we mean by direct and quadrature axis of synchronous machine?

Ans: A salient pole machine has two axes called as
(a) Direct axis (d-axis) or field pole axis and
(b) Quadrature axis (q-axis)

Direct axis is the axis, which passes through the poles whereas quadrature axis is the axis passing through the centres of the interpolar space.

Q. 38. Which test is useful for finding the values of X_d and X_q?

Ans: X_d and X_q are direct and quadrature axes reactances respectively and may be obtained by performing slip test on synchronous generator.

Q. 39. Why are the alternators connected in parallel ?

Ans: It is due to the reasons as given below:

(a) In case of breakdown/failure of any generator, there will be no interruption of power supply.

(b) Whenever one of the generators is taken out for maintenance purpose, remaining generators will supply the load continuously.

(c) During light load situations, some generators may be shut down and those remaining, operate at full load condition, so gives more efficient operation.

Q. 40. Which methods may be used for synchronization?

Ans: These methods are:

(a) Synchronizing lamps   and

(b) Synchroscope

Q. 41. What will happen, when two alternators are operating in parallel and the field current of the second alternator is reduced?

Ans: In this situation, the system terminal voltage is increased and the reactive power Q supplied by that alternator is increased, while the reactive power supplied by other alternator is decreased.

Q. 42. What are the necessary conditions for parallel operation of alternators?

Ans: The conditions must be:

(a) The phase sequence of bus bar must be same as that of incoming machine

(b) The frequency of the generated voltage of incoming machine must be equal to the frequency of the voltage of the live bus bar

(c) The bus bar voltage and the terminal voltage of the incoming machine must be equal

Q. 43. What is the meaning of stability limit in a synchronous machine?

Ans: In a synchronous machine, stability limit simply means the maximum power flow possible, when the synchronous machine is operating with stability.

Q. 44. What are the various reasons of hunting in a synchronous machine?

Ans: The reasons may be:
(a) A fault in the supply system
(b) Sudden change in the value of field current
(c) A load or drive containing harmonic torques, etc.

Q. 45. What is the purpose of damper bars in alternators?

Ans: The main purpose of bars is to damp-out rotor oscillations (hunting).

Q. 46. Name the various losses in synchronous machines?

Ans: The losses in synchronous machines are:
(a) Direct load loss (I^2R loss in armature winding)
(b) Field circuit loss (I^2R loss in field winding)
(c) No load rotational loss (friction and windage loss and open circuit core loss)
(d) Stray load loss (in iron and armature conductors)

Q. 47. Which parameter of load influences the armature reaction of an alternator?

Ans: Power factor of load.

7

Synchronous Motor

Q. 1. The synchronous motor is similar to (in construction):

(a) Squirrel cage induction motor

(b) Commutator motor

(c) Synchronous generator

(d) Double cage motor

Ans: **(c)**

Q. 2. Which one of the following machines has doubly excited magnetic system?

(a) Synchronous motor

(b) DC motor

(c) Slip ring induction motor

(d) Double cage motor

Ans: **(a)**

Q. 3. The parts of a 3-phase synchronous motor are:

(a) Laminated stator core

(b) Brushes and brush holder

(c) Rotating field structure

(d) All of the above

Ans: **(d)**

Q. 4. In synchronous motors, the power supplied to stator will be given by the expression:

(a) $\sqrt{3}\ I_L \cos\phi$ watts (b) $\sqrt{3}\ V_L I_L$ watts

(c) $\sqrt{3}\ V_L \cos\phi$ watts (d) $\sqrt{3}\ V_L I_L \cos\phi$ watts

Ans: **(d)**

Q. 5. Which one of the following torques may also be called as break-away torque?

(a) Running torque

(b) Pullout torque

(c) Starting torque

(d) Pull in torque

Ans: (c)

Q. 6. In a synchronous motor, running torque is determined by the:

(a) Output power

(b) Speed of the driven machine

(c) Both (a) and (b)

(d) None of these

Ans: (c)

Q. 7. Which of the following statements is not an advantage of synchronous motor?

(a) High operating efficiency

(b) Constant speed

(c) More sensitive to system disturbances

(d) None of the above

Ans: (c)

Q. 8. In a synchronous motor:

(a) Its rotor poles are excited by direct current and stator windings are given DC supply

(b) Its rotor poles are excited by direct current and the stator windings are connected to AC supply

(c) Its stator windings are connected to DC supply

(d) Its rotor poles may be excited by either AC or DC

Ans: (b)

Q. 9. Which one of the following statements is wrong?

(a) A synchronous motor is a doubly excited machine

(b) Damper windings are used for starting the synchronous motor

(c) Three-phase armature winding is on the rotor

(d) Synchronous motor runs either at synchronous speed or not at all

Ans: (c)

Q. 10. Whenever the load on a synchronous motor is increased, the:

(a) Torque angle δ reduces

(b) Armature current (I_a) increases

(c) Excitation voltage increases

(d) Armature current remains constant

Ans: (b)

Q. 11. In a synchronous motor, the pull out torque varies from:

(a) 2 to 4 times full load torque

(b) 3 to 7 times full load torque

(c) 1.5 to 3.5 times full load torque

(d) 2 to 2.5 times full load torque

Ans: (c)

Q. 12. When the load (on a synchronous motor) is increased, then its speed:

(a) Reduces

(b) Increases

(c) Remains unaltered

(d) Remains unaltered and additional load is supplied by shaft in relative position of the rotor w.r.t. stator rotating magnetic field

Ans: (d)

Q. 13. Which of the following statements is wrong?

(a) Synchronous motors are rated between 150 kW and 15 MW

(b) Synchronous motor may run at both lagging and leading power factors

(c) It can be used for power factor correction purposes

(d) None of these

Ans: (d)

Q. 14. In a synchronous motor, load angle depends mainly upon:

(a) Load   (b) Excitation

(c) Speed   (d) Applied voltage

Ans: (a)

Q. 15. The value of back emf in a synchronous motor:

(a) Is equal to zero

(b) May or may not be equal to zero

(c) May be either equal to or less than or more than supply voltage

(d) None of the above

Ans: (c)

Q. 16. A synchronous motor is able to supply increased mechanical load:

(a) By reduction in speed

(b) By shift in relative positions of the rotor and rotating magnetic field

(c) By increment in speed

(d) Both (b) and (c)

Ans: (b)

Q. 17. When huge mechanical load is applied to a synchronous motor, then:

(a) The rotor will pull out of synchronism

(b) After some time, the rotor will come to standstill

(c) Both (a) and (b)

(d) None of the above

Ans: (c)

Q. 18. In a synchronous motor, the flux produced by armature currents lags the respective armature currents by the angle:

(a) 70°   (b) 60°

(c) 90°   (d) 125°

Ans: (c)

Q. 19. With the change in field excitation of synchronous motor:

(a) The speed of motor will be affected

(b) The motor output will be affected

(c) The power factor will be affected

(d) Both (b) and (c)

Ans: **(c)**

Q. 20. For the overexcitation mode (synchronous motor):

(a) $E > V$ (b) $E < V$

(c) $E = V = 0$ (d) $E = V \neq 0$

Ans: **(a)**

Q. 21. With reduction in the value of DC excitation, synchronous motor draws:

(a) Less current at unity power factor

(b) More current from the supply mains at lower lagging power factor

(c) Less current at lagging power factor

(d) More current from the supply mains at leading power factor

Ans: **(b)**

Q. 22. In synchronous motor, the induced emf (E) is:

(a) Also known as back or counter emf

(b) In phase with the applied voltage

(c) Dependent upon the rotor flux per pole

(d) Both (a) and (c)

Ans: **(d)**

Q. 23. What must be the speed of a synchronous motor, when the load is reduced to half the full load (it runs at N_s rpm at full load condition)?

(a) $N_s/2$ rpm   (b) $N_s/4$ rpm

(c) $N_s/7$ rpm   (d) N_s rpm

Ans: **(d)**

Q. 24. The speed of a synchronous motor may be controlled by varying:

(a) Excitation of field winding

(b) Normal supply frequency only

(c) Applied voltage

(d) Both (b) and (c)

Ans: **(d)**

Q. 25. The excitation of a synchronous motor is adjusted to give unity power factor current, while it is operating at constant load. What must be the power factor, when the excitation is increased?

(a) Remains at unity   (b) Becomes zero

(c) Lagging nature   (d) Nothing can be said

Ans: **(a)**

Q. 26. In a synchronous machine, back emf (E_b) is:

(a) Set up in stator by the rotor flux

(b) Set up in rotor by stator flux

(c) Set up in stator by the stator flux

(d) Set up in rotor by rotor flux

Ans: **(a)**

Q. 27. Total value of (net) voltage in stator will be:

(a) Vector sum of applied voltage and back emf

(b) Vector difference of applied voltage and back emf

(c) Nearly zero

(d) Thrice the rated voltage

Ans: **(b)**

Q. 28. Whenever the synchronous motor is loaded, the rotor will fall back in phase by a greater value of angle α. This angle is known as:

(a) Coupling angle   (b) Power factor angle

(c) Load angle   (d) Both (a) and (c)

Ans: **(d)**

Q. 29. In synchronous motor, the mechanical power (P_m) developed in rotor will be:

(a) Back emf × rotor current

(b) Back emf × applied voltage

(c) Back emf × current flowing in armature × cosine of angle between I_a and E_b reversed

(d) Back emf × current flowing in armature × sine of angle between I_a and E_b reversed

Ans: (c)

Q. 30. In the rotor of synchronous motor, the copper loss is met by:

(a) Motor AC input only

(b) DC source used for rotor excitation

(c) AC source used for stator excitation only

(d) Both (a) and (b)

Ans: (b)

Q. 31. Which of the following expressions is correct?

(a) $V = (-E_b) + I_a Z_S$ (b) $V = (E_b) + I_a X_s$

(c) $V = (-E_b) + I_a X_S$ (d) $V = (E_b) + (-I_a X_s)$

Ans: (a)

Q. 32. In under and overexcited synchronous motors, the value of power factor reaches to which one of the following values with increment in load:

(a) 1

(b) 0.5

(c) A value between 0.2 to 0.4

(d) None of the above

Ans: (a)

Q. 33. A synchronous motor is started as an induction motor till it runs:

(a) 3 to 10% below the synchronous speed

(b) 2 to 5% below the synchronous speed

(c) 5 to 10% below the synchronous speed

(d) 10 to 20% below the synchronous speed

Ans: **(b)**

Q. 34. The power factor of synchronous machines may be controlled by the variation of:

(a) Field excitation

(b) Armature current

(c) Impedance of stator side

(d) Supply frequency

Ans: **(a)**

Q. 35. In a synchronous motor, some mechanical power is developed, even if the field is unexcited. It should be a:

(a) Cylindrical rotor machine

(b) Salient pole machine

(c) Nothing can be said

(d) Any one of the above

Ans: **(b)**

Q. 36. The V-curve of a synchronous motor is drawn between:

(a) Armature current and field current

(b) Field current and supply voltage

(c) Armature current and power factor

(d) Power factor and field current

Ans: **(a)**

Q. 37. The speed of a synchronous motor may be varied by:

(a) Variation of supply voltage only

(b) Variation of frequency of supply source

(c) Variation of field current

(d) Any one of the above

Ans: **(b)**

Q. 38. Which one of the followings is the advantage of synchronous motor?

(a) Constant speed

(b) Good efficiency

(c) Capability of being operated under a wide range of power factor (both lagging and leading)

(d) All of these

Ans: **(d)**

Q. 39. Maximum power will be delievered by synchronous motor, when:

(a) The value of input power factor is 0.5

(b) The value of load angle is 90°

(c) The load angle = internal angle (θ)

(d) The value of input power factor is 0.2

Ans: **(c)**

Q. 40. A pony motor is a:

(a) Small induction motor

(b) DC compound wound motor

(c) Double cage synchronous motor

(d) DC shunt motor

Ans: **(a)**

Q. 41. A three-phase synchronous motor is mostly used for:

(a) High torque loads

(b) Control of voltage at the end of transmission line

(c) The improvement of power factor

(d) Both (b) and (c)

Ans: **(d)**

Q. 42. If suddenly one of the phases of a 3-phase synchronous motor is short circuited, then the motor will:

(a) Run at 1/3rd of synchronous speed

(b) Take less than the rated load

(c) Run at 1/2 synchronous speed

(d) Nothing can be said perfectly

Ans: **(b)**

Q. 43. A synchronous condenser is:

(a) An over-excited synchronous motor driving mechanical load

(b) An over-excited synchronous motor without mechanical load

(c) An under-excited synchronous motor driving mechanical load

(d) An under-excited synchronous motor without mechanical load

Ans: **(b)**

Q. 44. Which of the following motors is capable of operating at both power (lagging and leading) factors?

(a) Synchronous motor

(b) Double s/c motor

(c) Single-phase induction motor

(d) Universal motor

Ans: **(a)**

MISCELLANEOUS

Q. 1. In a synchronous machine, when the axis of field flux coincides with that of armature flux, the machine:

(a) Is said to be floating

(b) Will operate as an induction motor

(c) Will operate as a synchronous generator

(d) Will operate as an induction generator

Ans: **(a)**

Q. 2. A synchronous motor runs at synchronous speed only because of:

(a) The magnetic locking between stator and rotor poles

(b) Bio-Savart's law

(c) Lenz's law

(d) Auxiliary winding

Ans: **(a)**

Q. 3. Whenever load on a synchronous motor (running with normal excitation) is increased, armature current drawn by it increases, because:

(a) E_R (net resultant voltage) in armature is increased

(b) Power factor is increased

(c) E_R in armature is decreased

(d) Any one of the above

Ans: (a)

Q. 4. When the main field current of salient pole synchronous motor, fed from an infinite bus and running at no load is reduced to zero, it would:

(a) Continue running at synchronous speed

(b) Not run at all

(c) Give half the rated output

(d) Run at nearly 1/7th of synchronous speed

Ans: (a)

Q. 5. Maximum power developed in a synchronous motor will be:

(a) $P_{max} = VE_f/Z_s + E_f^2 R_a/Z_s^2$ (b) $P_{max} = VE_f/Z_s - E_f^2 R_a/Z_s^2$

(c) $P_{max} = V^2 E_f/Z_s$ (d) $P_{max} = \dfrac{VE_f^2}{Z_s} R_a$

Ans: (b)

Q. 6. Which one of the followings is the effect of hunting?

(a) In rotor shaft, great mechanical stress may be developed

(b) The final temperature of machine may increase

(c) Possibility of resonance may increase

(d) All of the above

Ans: (d)

Q. 7. When load is increased, synchronous motor draws more armature current as:

(a) The rotor pole falls back relative to stator pole

(b) Stator poles fall back relative to rotor poles

(c) Stator pole moves forward relative to rotor pole

(d) None of these

Ans: **(a)**

Q. 8. The interaction torque depends upon:

(a) Stator and rotor field strength

(b) Torque angle

(c) Power factor only

(d) Both (a) and (b)

Ans: **(d)**

Q. 9. What will be the speed of synchronous motor, if the following data is halved for a 3-phase; 4 pole; 50 Hz motor?

  (i) Pole number   **(ii) Load torque**

 (iii) The frequency

(a) 2500 rpm   (b) 1700 rpm

(c) 2000 rpm   (d) 1500 rpm

Ans: **(d)**

Q. 10. When damper winding is used in a 3-phase synchronous motor, then it starts as a:

(a) 3φ induction generator

(b) 3φ induction motor

(c) 1φ alternator

(d) DC compound wound motor

Ans: **(b)**

Q. 11. For synchronous motors, the standard full load power factor ratings will be:

(a) Unity or 0.8 lead   (b) 0.5 lead

(c) 0.5 lag   (d) Nearly zero

Ans: **(a)**

Q. 12. In a synchronous motor, the field circuit is kept in ON condition but the stator winding is disconnected from the supply. In this condition, the flux per pole was

25 mWb. After some time period, the stator is connected to the rated supply with the field excitation unchanged, then the flux per pole in the machine is found to be nearly 20 mWb, while the motor is running on "no load". (Assume no load losses to be negligible). In this condition, the no load current taken by the motor from the supply side:

(a) Leads the supply voltage

(b) Will be zero

(c) Will be nearly 2 A

(d) Will be nearly 0.2 A

Ans: (a)

Q. 13. What will be the value of kilowatt ampere rating of synchronous condenser, if it is installed to supply 2400 kVA, 0.65 power factor load in a plant to unity?

(a) 1825 MVA   (b) 1825 kVA

(c) 1700 MVA   (d) 1700 kVA

Ans: (b)

Q. 14. A synchronous motor and 3φ SRIM are mechanically coupled and operate on the 3φ, 50 Hz supply system. The data for the machines is as follows:

(i) 3φ synchronous motor ——————— $p = 6, f = 50$ Hz

(ii) 3φ slip ring induction motor —— $p = 8, f = 50$ Hz

In this condition, what must be the value of frequency of the voltage produced across any two rings, if they are kept open circuited?

(a) 50/3 Hz   (b) 50 Hz

(c) 32 Hz   (d) 20.5 Hz

Ans: (a)

Q. 15. A synchronous motor is allowed to operate at no load condition at unity power factor. When the field current is increased, then:

(a) The power factor will be of lagging nature

(b) Power factor will be of leading nature

(c) The current will increase

(d) Both (b) and (c)

Ans: **(d)**

Q. 16. A 3-phase synchronous motor is connected to the supply at constant terminal voltage (V_t) and rated frequency (consider the losses to be negligible). Now when the motor is gradually loaded to its rated power, adjusting its excitation to obtain, say 0.8 lagging power factor operation, then the induced emf phasor would:

(a) Be greater than V_t phasor

(b) Lag V_t phasor

(c) Be less than V_t phasor

(d) Both (a) and (b)

Ans: **(d)**

Q. 17. A synchronous motor is running at a load angle of 20° at rated frequency with negligible armature resistance. What will be the value of new load angle, if:

(1) Other parameters are kept constant

(2) The supply frequency is raised by 10%

(a) 30° (b) 20°

(c) 25° (d) 35°

Ans: **(b)**

Q. 18. A synchronous motor is working at half full load and is being fed from an infinite bus. If an increase in the field current causes a reduction in the armature current, then the motor is:

(a) Absorbing reactive power from the bus at lagging power factor

(b) Delievering reactive power to the bus at leading power factor

(c) Both (a) and (b)

(d) Nothing can be said

Ans: **(a)**

Q. 19. **A synchronous motor (connected to an infinite bus) is operating at a certain load angle. When a sudden increment in load angle takes place, then the synchronising power flows from:**

(a) Generator to infinite bus

(b) Infinite bus to motor

(c) Infinite bus to generator

(d) Both (a) and (b)

Ans: (d)

Q. 20. **A synchronous motor (salient pole) is running at no load in clockwise direction. When its field current is reduced to approximately zero and then reversed, then:**

(a) The rotor slips by one pole pitch

(b) Rotor continues running at synchronous speed

(c) Rotor slips by 2/3rd pole pitch

(d) Both (a) and (b)

Ans: (d)

FILL IN THE BLANKS

1. The construction of a polyphase synchronous motor is same as that of a ....

Ans: Synchronous generator

2. The rotor of the synchronous motor usually has .... poles connected to give alternate polarity.

Ans: Salient field

3. Synchronous motors are manufactured in ratings ranging from .... kW to .... MW.

Ans: 150; 15

4. Synchronous motor has no .... torque.

Ans: Self starting

5. A synchronous motor is basically .... excited machine.

Ans: Doubly

6. Synchronous motors may be operated under wide range of power factors both .... and ....

Ans: Lagging; Leading

7. A synchronous motor is started as .... motor till it runs 2% to 5% below the synchronous speed.

Ans: Induction

8. .... torque is the maximum value of torque which a synchronous motor can develop at rated voltage and frequency.

Ans: Pull out

9. Synchronous motors run at .... speed, regardless of load.

Ans: Constant synchronous

10. The value of torque angle δ is increased, when the load on synchronous motor is ....

Ans: Increased

11. The value of pull-out torque varies from .... to .... times full load torque.

Ans: 1.5; 3.5

12. The synchronous motor works with .... power factor at the minimum armature current.

Ans: Unity

13. V-curve is plotted between .... and .... for synchronous motors.

Ans: Field current; Armature current

14. The 'V' shaped curve, drawn for synchronous motors may be useful in adjusting .... current.

Ans: Field

15. Inverted V-curve may be plotted between field current and ....

Ans: Power factor

16. So as to draw V-curves, the motor is run from constant .... and .... bus bars.

Ans: Voltage; Constant frequency

17. A synchronous motor may be started by means of an external prime mover or .... windings.

Ans: Damper

18. Damper windings consist of heavy .... bars inserted in slots of .... of rotor.

Ans: Copper; Pole faces

19. Initially, synchronous motor starts as a 3-phase ....

Ans: Induction motor

20. In synchronous motors, the flux produced by armature currents .... respective armature currents by 90°.

Ans: Lags

21. When $E > V$, then the motor is said to be operating in .... mode.

Ans: Over-excited

22. In synchronous motors, DC supply is fed to .... winding.

Ans: Field

23. Excitation voltage may also be known as .... or .... emf.

Ans: Back; Counter

24. Effective value of resistance is usually .... to .... of X_s.

Ans: 1/100; 1/10

25. The value of power supplied to stator of synchronous motor is expressed as .... watts.

Ans: $\sqrt{3}\ V_L I_L \cos\theta$

26. Break away torque in synchronous motor may also be known as .... torque.

Ans: Starting

27. The circle diagram for synchronous motors is an extension of .... diagram.

Ans: Synchronous impedance

28. The constant power circle is not of great practical interest because the value of radius is ....

Ans: Large

29. When a synchronous motor is operated at no load condition, it still takes power for .... losses.

Ans: Rotational

30. The phenomenon of oscillation of the rotor is about its final equilibrium position is known as ....

Ans: Hunting

31. Hunting phenomenon may also be known as ....

Ans: Phase swinging

32. Hunting phenomenon .... the possibility of resonance.

Ans: Increases

33. Synchronous motor is .... efficient compared to induction motor of the same output and voltage rating.

Ans: More

34. Synchronous motor is .... compared to induction motor.

Ans: Costlier

35. Synchronous condenser may also be called as synchronous ....

Ans: Phase modifier

36. Rotational losses in the electrical machine is the sum of friction, windage and .... losses.

Ans: Iron

37. When a synchronous condenser carries full rated leading kVA, the current leads the voltage by .... degrees.

Ans: 88 to 89

38. So as to avoid high starting current, the synchronous motor is usually started as a .... motor.

Ans: Slip ring induction

39. At starting moment, the field windings must be shorted through a suitable .... resistance.

Ans: Field

40. In a synchronous motor, the value of starting torque may be increased by increasing the resistance of .... winding.

Ans: Rotor

41. The speed of synchronous motor depends upon .... and supply frequency.

Ans: Total number of poles

42. In synchronous motor, the electromagnetic power varies linearly with ....

Ans: Voltage

43. In synchronous motors, the cost per kW output is .... compared to induction motors.

Ans: Higher

44. These motors may be used in parallel to the bus bars so as to improve the .... factor.

Ans: Power

45. Synchronous motors may be used to regulate .... at the end of transmission lines.

Ans: Voltage

46. These motors may also be employed for loads, where .... speed is needed.

Ans: Constant

TRUE/FALSE

1. A three-phase synchronous machine is a singly excited machine.

Ans: False

2. Most of the synchronous motors are of salient-pole type.

Ans: True

3. The stator core of a synchronous motor is similar to that of an alternator.

Ans: True

4. The rotor of synchronous motor has non-salient field poles.

Ans: False

5. When the excitation is 100%, the armature current (I) leads the resultant voltage.

Ans: False

6. The kilowatt rating of an exciter for a synchronous motor is approximately 7% of the kVA rating of the synchronous motor.

Ans: False

7. The reduced excitation to synchronous motor causes lagging power factor.

Ans: True

8. At unity power factor, maximum current is taken by the synchronous motor.

Ans: False

9. The phasor diagram of a synchronous motor is same as that of an alternator.

Ans: False

10. In synchronous motors, the phasor of applied voltage (V) lags the phasor of excitation voltage.

Ans: False

11. Pull out torque may also be known as running torque.

Ans: False

12. In a salient pole synchronous motor, the air gap is not uniform.

Ans: True

13. Salient pole synchronous motor has two axis namely d-axis and q-axis.

Ans: True

14. Rigidity factor may also be called as stability factor.

Ans: True

15. In synchronous motors, rotational losses are very low.

Ans: False

16. The speed of synchronous motor is independent of load.

Ans: True

17. The air gap flux is the difference of fluxes due to rotor current and stator current.

Ans: False

18. Locked rotor torque is provided by the stator windings.

Ans: **True**

19. In overexcited mode, the synchronous motor has a leading current.

Ans: **True**

20. When the load is increased, the phase angle Ø increases in the lagging direction.

Ans: **True**

21. The power factor of a synchronous motor can be controlled by variation of armature current.

Ans: **False**

22. V-curve of a synchronous motor is plotted between field current (I_f) and armature current (I_a).

Ans: **True**

23. Inverted V-curve may be plotted between armature current and power factor.

Ans: **False**

24. A damper winding consists of light weight aluminium bars.

Ans: **False**

25. Phase swinging phenomenon is also known as hunting.

Ans: **True**

26. Hunting occurs in both synchronous motors and 1φ induction motors.

Ans: **False**

27. Sudden changes in the value of armature current may cause phase swinging.

Ans: **False**

28. Due to hunting, large mechanical stresses may develop in the rotor shaft.

Ans: **True**

29. A synchronous motor may be used for driving mechanical loads only.

Ans: **False**

30. When synchronous motor is operated at full load with over excitation, it takes a current which leads the voltage by 70°.

Ans: **False**

31. Synchronous condenser compensates the lagging current drawn by induction motors so as to improve the overall power factor of plant.

Ans: **True**

32. The changes in applied voltage do not affect synchronous motor torque as much as they affect induction motor torque.

Ans: **True**

33. Synchronous motors are well suited for high voltage services.

Ans: **True**

34. Synchronous motor may be used in reciprocating compressor drives.

Ans: **True**

35. The maximum value of torque angle in a synchronous motor is 90° electrical.

Ans: **True**

36. When the field of synchronous motor is under-excited, the power factor will be nearly unity.

Ans: **False**

37. When the field excitation is excessive, a synchronous motor may fail to start.

Ans: **True**

38. Large static friction may be the cause of faulty starting of a synchronous motor.

Ans: **True**

VIVA VOCE QUESTIONS

Q. 1. What do you mean by synchronous motor?

Ans: It is a machine that converts AC electric power to mechanical power at a constant speed called synchronous speed.

Q. 2. How can we change the speed of a synchronous motor?

Ans: The speed of a synchronous motor may be changed by either varying the number of poles or supply frequency.

Q. 3. Why the 3-phase synchronous motor runs at synchronous speed?

Ans: Due to the interlocking action between stator and rotor fields of motor.

Q. 4. Does the change in excitation affect the speed of a synchronous motor?

Ans: No.

Q. 5. Which type of AC motor may be used for power factor correction purposes?

Ans: Synchronous motor.

Q. 6. What will be the mode of excitation of synchronous motor if $E_b < V$?

Ans: Under-excited mode.

Q. 7. What do you mean by excitation or counter emf in a synchronous motor?

Ans: In synchronous motor, a voltage is induced in the stator winding due to field rotating at synchronous speed. This induced emf is known as counter emf (opposes the applied voltage V to the stator).

Q. 8. What do you mean by break-away torque in a synchronous motor?

Ans: It is also known as starting torque. Actually it is related to the ability of the motor to accelerate the load.

Q. 9. What is pull out torque?

Ans: It is the maximum torque, which will be developed by synchronous motor without pulling out of step.

Q. 10. What is 'V-curve' for a synchronous motor?

Ans: V-curve is a curve drawn between field current (X-axis) and armature current (Y-axis) of synchronous motor.

Q. 11. What must be the value of power factor, when the armature current becomes minimum?

Ans: It must be 1, i.e. unity power factor.

Q. 12. What are the main characteristic features of a synchronous motor?

Ans: The main characteristic features include:

(a) Synchronous motor will run at a speed, called as synchronous speed, given by $N_s = 120\,f/p$

(b) It can be operated at both lagging and leading power factors.

(c) Synchronous motor is inherently not self starting so it is required to run up to synchronous speed by some means before connecting AC supply source.

Q. 13. What do you mean by synchronous condenser?

Ans: When the synchronous motor is operated in over-excited mode at no load condition so that the current drawn by it leads the voltage by nearly 90°, then the machine is called as synchronous condenser.

Q. 14. What is the most important advantage of synchronous condenser?

Ans: The advantage is that the power factor can very easily be controlled by variation of field excitation.

Q. 15. What are the starting methods of synchronous motor?

Ans: These methods include:

(a) By using an AC motor

(b) By means of damper grids in the pole faces

(c) From DC source

Q. 16. What are the applications of synchronous motor?

Ans: The application are as follows:

(a) These motors are used to regulate the voltage at the end of transmission lines

(b) It may be used as drives for blowers, fans, line shafts, compressors, etc.

(c) These may also be used for improving power factor of the system

8

Three-Phase Induction Motor

Q. 1. An induction motor works on:

(a) Alternating current supply

(b) Direct current supply

(c) Both (a) and (b)

(d) None of these

Ans: **(a)**

Q. 2. Three-phase induction motors may be classified according to:

(a) Principle of operation

(b) Their speed

(c) Their structural features

(d) All of the above

Ans: **(d)**

Q. 3. A 3-phase induction motor may also be called as:

(a) Conduction motor   (b) Rotating transformer

(c) Stationary transformer   (d) None of these

Ans: **(b)**

Q. 4. In a three-phase induction motor, the core of stator is laminated. It's purpose is to reduce:

(a) Rotational losses

(b) Iron losses

(c) Eddy current losses

(d) Friction and windage losses

Ans: **(c)**

Q. 5. In induction motors:

(a) Stator winding is open circuited and rotor winding is connected to AC supply

(b) Stator winding is connected to AC supply source and rotor winding is short circuited

(c) Stator winding is connected to DC supply and rotor winding is open circuited

(d) Both stator and rotor windings are given AC supply

Ans: (b)

Q. 6. The purpose of using stator frame is to:

(a) Protect brushes and slip rings

(b) Protect the entire body of machine

(c) Hold the armature stampings in position

(d) Ventilate the complete armature

Ans: (c)

Q. 7. The results of skewing in a 3-phase induction motor include:

(a) Increased slip for a given torque

(b) Reduction in rotor resistance and noise

(c) Increased rotor resistance due to increased length of rotor bars

(d) Both (a) and (c)

Ans: (d)

Q. 8. In 3-phase induction motors, slip rings are made of:

(a) Light aluminium

(b) Copper alloy

(c) High quality phosphor-bronze

(d) Good quality alloy

Ans: (c)

Q. 9. The induction motor frame is made of:

(a) Mild steel

(b) Silicon steel

(c) Closed grained cast iron

(d) Copper

Ans: (c)

Q. 10. The stator and rotor cores of an induction motor are made up of magnetic material:

(a) To increase the magnetizing current

(b) To make the parts strong

(c) To keep the overall cost low

(d) To reduce the magnetizing current

Ans: (d)

Q. 11. In a 3-phase induction motor, the value of magnetizing current is much greater compared to an equivalent transformer, because:

(a) Magnetic material is used for core construction

(b) Of the presence of air gap between stator and rotor in an induction motor

(c) The size of induction machine is smaller than an equivalent transformer

(d) None of the above

Ans: (b)

Q. 12. Now-a-days, die-cast aluminium rotor is used, because aluminium is:

(a) Cheaper and lighter than copper

(b) More ductile compared to copper

(c) Easy to cast due to low melting point and easily available

(d) Both (a) and (c)

Ans: (c)

Q. 13. The rotor of an induction motor never reaches synchronous speed, because:

(a) The relative speed between the rotating flux and the rotor will be zero and so torque will also be zero

(b) The relative speed between flux (stator) and rotor will be maximum and hence torque will be zero

(c) The relative speed between stator flux and rotor will be minimum so torque will be maximum

(d) None of these

Ans: **(a)**

Q. 14. The balls are made up of ...... in ball bearings used in induction motor:

(a) Cast iron

(b) Carbon-chrome steel

(c) Stainless steel

(d) Pure copper

Ans: **(b)**

Q. 15. In induction motors, the thickness of punching (silicon-steel) varies from:

(a) 0.30 mm to 0.75 mm

(b) 0.40 mm to 0.60 mm

(c) 0.35 mm to 0.65 mm

(d) 0.35 mm to 0.70 mm

Ans: **(c)**

Q. 16. In a 3-phase induction motor, iron loss mainly occurs in:

(a) Stator and rotor teeth

(b) Rotor winding

(c) Stator core and stator teeth

(d) None of these

Ans: **(c)**

Q. 17. The air gap flux density is kept low so as to:

(a) Improve power factor

(b) Improve output

(c) Eliminate iron losses

(d) Reduce copper losses

Ans: **(a)**

Q. 18. The rotor of a 3-phase induction motor rotates in the same direction as that of stator rotating field. It may be explained by:

(a) Bio-Savart's law   (b) Fleming's left hand rule

(c) Lenz's law   (d) Newton's law

Ans: (c)

Q. 19. The direction of the rotating magnetic field developed by the stator ampere-turns of a three-phase induction motor changes, if:

(a) The applied voltage is reduced

(b) Any one phase is disconnected from supply

(c) The sequence of supply to the stator terminals is changed

(d) None of these

Ans: (c)

Q. 20. Which one of the following statements about wound rotor is incorrect?

(a) High starting torque

(b) High starting torque and low starting current

(c) Additional resistance may be connected in the rotor circuit for speed controlling purpose

(d) Low starting torque

Ans: (d)

Q. 21. Which one of the following options is not correct regarding cage rotor?

(a) Lesser maintenance

(b) Higher efficiency and moderate power factor

(c) Robust construction

(d) Risk of sparking is more

Ans: (d)

Q. 22. The rotor of an induction motor runs at:

(a) Synchronous speed

(b) Above synchronous speed

(c) Sometimes above and sometimes below synchronous speed

(d) Below synchronous speed

Ans: **(d)**

Q. 23. A coil is rotating in a magnetic field and its two ends are connected to two slip rings. What must be the nature of current available in the external circuit?

(a) Alternating current

(b) Direct current

(c) Mixture of (a) and (b)

(d) None of these

Ans: **(a)**

Q. 24. An induction motor may also be called as:

(a) Asynchronous motor   (b) Rotating transformer

(c) Conduction motor   (d) Both (a) and (b)

Ans: **(d)**

Q. 25. Which of the following motors is widely preferred at industrial level?

(a) Commutator motor   (b) 3-phase induction motor

(c) Synchronous motor   (d) Conduction motor

Ans: **(b)**

Q. 26. A 3-phase induction motor having wound rotor is known as:

(a) Slip ring induction motor

(b) Squirrel cage induction motor

(c) Double squirrel cage induction motor

(d) Capacitor start motor

Ans: **(a)**

Q. 27. What will happen, if the air gap of an induction motor is increased?

(a) Power factor will decrease

(b) Magnetising current of rotor will decrease

(c) Power factor will increase for some movement and then decreases continuously

(d) Power factor will increase

Ans: **(b)**

Q. 28. In an induction motor, percentage slip depends on:

(a) Motor copper losses

(b) Supply voltage and frequency

(c) Frequency only

(d) None of these

Ans: **(a)**

Q. 29. The purpose of using large number of slots in induction motor is to:

(a) Reduce motor size

(b) Increase power factor and to reduce magnetizing current

(c) Increase the value of air gap flux

(d) Provide better overload capacity

Ans: **(d)**

Q. 30. When air gap length in induction motors is large, it will:

(a) Reduce humming noise of motor

(b) Increase humming noise of motor

(c) Increase overload capacity of the motor

(d) Reduce the overload capacity of motor

Ans: **(c)**

Q. 31. Whenever 3-phase induction motor is allowed to run, the frequency of rotor current will be:

(a) Somewhat less compared to stator frequency

(b) More than stator frequency

(c) Less than stator (supply) frequency

(d) None of these

Ans: **(c)**

Q. 32. The operating principle of a 3-phase induction motor closely matches to that of a:

(a) Synchronous generator

(b) Commutator motor

(c) Two winding transformer with its secondary short circuited

(d) Three winding transformer

Ans: (c)

Q. 33. If the rotor circuit of 3-phase induction motor is open, then it will:

(a) Not run at all

(b) Make loud noise

(c) Run in normal mode

(d) Run at fast speed

Ans: (a)

Q. 34. A 3-phase induction motor never runs at synchronous speed. If it did so, then:

(a) The relative speed between rotating flux and rotor will be zero

(b) It will produce a torque, which is lesser in magnitude

(c) Emf induced in the rotor circuit and the torque developed will be zero

(d) Both (a) and (c)

Ans: (d)

Q. 35. The slip of induction motor is given as:

(a) $(N - N_s)/8$ (b) $(N_s - N)/N$

(c) $(N - N_s)/N^2$ (d) $(N_s - N)/N_s$

Ans: (d)

Q. 36. The actual speed of rotor of an induction motor is given as:

(a) $0.80\ N_s$ (b) $(1 - S)N_s$

(c) $S\ N_s$ (d) N_s

Ans: (b)

Q. 37. In a squirrel cage induction motor, the data is:
$f = 60$ Hz, $P = 12$, Full load slip = 5%
What will be its full load speed?

(a) 570 rpm   (b) 700 rpm
(c) 650 rpm   (d) 550 rpm

Ans: **(a)**

Q. 38. The speed of rotating field (due to rotor currents) relative to rotor surface in an induction motor is:

(a) $(1 - S)\, N_s$ (b) $(2 - S)\, N_s$
(c) N_s (d) SN_s

Ans: **(d)**

Q. 39. What will be the rotor speed in an induction motor, if a centre–zero ammeter is providing 30 oscillations per minute? Number of poles are 6 and frequency is 50 Hz.

(a) 990 rpm   (b) 1020 rpm
(c) 1500 rpm   (d) 970 rpm

Ans: **(a)**

Q. 40. A 400 volts, 50 Hz, 3-phase star connected induction motor is running at a fixed speed. However, if the stator to rotor turn ratio is 2, then what must be the standstill rotor induced emf per phase?

(a) 115.5 volts
(b) 520 volts
(c) 235 volts
(d) 312.5 volts

Ans: **(a)**

Q. 41. In induction motors, the phase reactance is ...... compared to phase resistance:

(a) Very high
(b) Very small
(c) Slightly high
(d) Slightly small

Ans: **(a)**

Q. 42. The torque developed in a 3-phase induction motor is given as:

(a) $E_2\cos\theta_2$ (b) $E_2 I_2\cos\theta_2$

(c) $E_2\cos\theta_2/I_2$ (d) $E_2 I_2$

Ans: **(b)**

Q. 43. The relation between the torque developed and the supply voltage will be:

(a) T proportional to V (b) T proportional to $8V^2$

(c) T proportional to $\dfrac{1}{\sqrt{V}}$ (d) T proportional to V^2

Ans: **(d)**

Q. 44. In squirrel cage induction motor, the starting torque is:

(a) Slightly more than the full load torque

(b) Very low

(c) Slightly low compared to full load torque

(d) Approximately zero

Ans: **(a)**

Q. 45. The starting torque will be maximum only, when:

(a) $(R_2)^2 = X_2$ (b) $R_2 = (X_2)^2$

(c) $R_2 = X_2$ (d) $R_2 = \sqrt{(X_2)}$

Ans: **(c)**

Q. 46. The starting torque will be high for a:

(a) Slip ring induction motor

(b) Squirrel cage induction motor

(c) Motor, which is being operated on AC supply

(d) None of these

Ans: **(a)**

Q. 47. The torque of an induction motor is:

(a) Inversely proportional to slip

(b) Directly proportional to slip

(c) Proportional to applied voltage

(d) Independent of power factor

Ans: **(b)**

Q. 48. By increasing ......, the starting torque of a 3-phase induction motor can be raised:

(a) Current   (b) Applied voltage

(c) Slip   (d) All of the above

Ans: **(d)**

Q. 49. The emf (induced) in the rotor of an induction motor is proportional to:

(a) Slip

(b) Relative velocity between flux and rotor conductors

(c) Applied stator voltage

(d) Both (b) and (c)

Ans: **(d)**

Q. 50. In a 3-phase squirrel cage induction motor, the starting torque will be:

(a) Same as full load torque

(b) Nearly 1.5 times the full load torque

(c) Zero

(d) None of the above

Ans: **(b)**

Q. 51. An induction motor possesses rotor resistance of 0.003 ohms/phase. When it is raised to 0.006 ohms/phase, the maximum torque will:

(a) Increase by 200%

(b) Reduce by 10%

(c) Remain unchanged

(d) Reduce by 35%

Ans: **(c)**

Q. 52. Whenever rotor resistance is increased, then the maximum torque occurs at:

(a) Lower speed   (b) $1/7$th N_s

(c) $1/5$th N_s (d) Very high speed

Ans: **(a)**

Q. 53. Which of the following statements is not correct?

(a) Slip = Rotor copper loss/rotor input power

(b) Slip = $(N_s - N)/N_s$

(c) Rotor input = input to motor – stator copper and iron losses

(d) Rotor copper loss/rotor output power = Slip

Ans: **(d)**

Q. 54. The torque depends upon:

(a) Rotor copper losses and rotor input

(b) Slip

(c) Both (a) and (b)

(d) None of the above

Ans: **(c)**

Q. 55. The relation between air gap power (P_g), mechanical power developed and the slip will be:

(a) $P_m = (1 - S) P_g$ (b) $P_m = P_g/1 - S$

(c) $P_g = (2 - S)/P_m$ (d) $P_m = (1 - S)/P_g$

Ans: **(a)**

Q. 56. The power factor of a lightly loaded induction motor is low because:

(a) The current drawn is largely a magnetizing current due to air gap

(b) The value of current at light load will be low

(c) Current drawn is mainly capacitive current

(d) Current drawn is very high due to large impedance

Ans: **(a)**

Q. 57. Whenever the load is increased from light load in a 3-phase induction motor, then:

(a) Stator power factor increases

(b) Rotor power factor decreases

(c) Both stator and rotor power factor increases

(d) Both (a) and (b)

Ans: **(d)**

Q. 58. When additional resistance is inserted in the stator of an induction motor, then the value of starting torque will:

(a) Decrease

(b) Increase

(c) Remain unaltered

(d) Decrease first and then increases continuously

Ans: **(a)**

Q. 59. Which of the following tests may be performed on 3-phase induction motor?

(a) No load test   (b) Blocked rotor test

(c) stator resistance test   (d) All of the above

Ans: **(d)**

Q. 60. Which of the following tests is also called as short circuit test?

(a) Blocked rotor test   (b) Open circuit test

(c) Stator resistance test   (d) Both (b) and (c)

Ans: **(a)**

Q. 61. The purpose of short circuit test on an induction motor is to find:

(a) Power factor at short circuit

(b) Equivalent values of resistance and reactance

(c) Short circuit current at rated voltage

(d) All of the above

Ans: **(d)**

Q. 62. The total power input in no load test, performed on a three-phase induction motor is nearly equal to:

(a) Iron loss in the core

(b) Dissipation loss in primary winding

(c) The sum of I^2R loss and hysteresis loss in the core

(d) Eddy current loss in the core

Ans: **(a)**

Q. 63. During no load test, wattmeter reading gives:

(a) Stator I^2R loss

(b) Stator core loss

(c) Friction and windage loss

(d) All of the above

Ans: (d)

Q. 64. In a 3-phase induction motor, the equivalent circuit has variable resistance as mechanical load and is given as:

(a) $r_2 (1/S - 1)$ (b) $r_2/S - 1$

(c) $r_2 (1 - S)$ (d) r_2/S

Ans: (a)

Q. 65. The circle diagram of an induction motor can be drawn by using the data obtained from:

(a) No load test   (b) Short circuit test

(c) Stator resistance test   (d) All of the above

Ans: (d)

Q. 66. Which of the following quantities can not be represented by circle diagram?

(a) Maximum output   (b) Maximum torque

(c) Rotor input   (d) None of the above

Ans: (d)

Q. 67. Induction motors, when directly switched, take .... times their full load current:

(a) 5 to 7   (b) 7 to 10

(c) 3 to 5   (d) 4 to 6

Ans: (a)

Q. 68. Which of the following methods is used for starting slip ring induction motor?

(a) Star-delta switch arrangement

(b) Auto transformer

(c) Primary resistor method

(d) Rotor rheostat method

Ans: (d)

Q. 69. Auto starters or compensators may be used for:

(a) Star connected motors   (b) Delta connected motors

(c) Both (a) and (b)   (d) None of these

Ans: (c)

Q. 70. What must be the value of tapping on an auto transformer (required for a squirrel cage motor) to start the motor against 1/4th of full load torque. Also given that $I_{sc} = 4$, I_f and $S_f = 3\%$.

(a) 72.2%   (b) 25%

(c) 66.6%   (d) 53.25%

Ans: (a)

Q. 71. While starting a three-phase induction motor, a star-delta starter is used to:

(a) Reduce the starting current to a safe value

(b) Obtain good power factor

(c) Improve the magnetizing current

(d) None of these

Ans: (a)

Q. 72. A 3-phase star connected squirrel cage induction motor is used to drive milling machine. It has following specifications:

5 HP, three-phase, 400 V

At starting, the starting current will be

(a) 40 A   (b) 120 A

(c) 270 A   (d) 35 A

Ans: (a)

Q. 73. What will be the ratio of starting torque to full load torque in an induction motor, when it is started by means of star-delta switch. The specifications are as follows: $I_{sc} = 6$, I_f and $S_f = 4\%$.

(a) 0.48   (b) 0.32

(c) 0.55   (d) 0.42

Ans: (a)

Q. 74. In 3-phase induction motors, star delta starter may be used, because it:

(a) Provides permissible minimum starting current

(b) Prevents heating of motor winding

(c) Is regulated by electricity authority

(d) All of the above

Ans: **(d)**

Q. 75. Which of the following methods is inferior in view of the poor starting torque per ampere of the line current drawn?

(a) Series-inductor method of starting

(b) Auto compensator method

(c) Star-delta switch

(d) DOL starting

Ans: **(a)**

Q. 76. "Rotor resistance starting" is preferred for slip ring induction motors, because it:

(a) Improves starting torque and starting power factor

(b) Reduces starting current

(c) Reduces the maximum torque

(d) Both (a) and (b)

Ans: **(d)**

Q. 77. "Crawling phenomenon" occurs in induction motors due to:

(a) Harmonic induction torques

(b) Harmonic synchronous torques

(c) Both (a) and (b)

(d) None of the above

Ans: **(c)**

Q. 78. "Crawling" of induction motor simply means that it is running at speed of:

(a) One-seventh of synchronous speed

(b) Two-third of synchronous speed

(c) One-fourth of synchronous speed

(d) One-fifth of synchronous speed

Ans: **(a)**

Q. 79. The phenomenon of (magnetic) locking between stator and rotor teeth in induction motor is called as:

(a) Cogging   (b) Teeth locking

(c) Magnetic locking   (d) All of the above

Ans: **(d)**

Q. 80. The teeth locking mainly occurs due to:

(a) Harmonic synchronous torques only

(b) Harmonic induction torques only

(c) Both (a) and (b)

(d) Vibration torques

Ans: **(a)**

Q. 81. In double squirrel cage motors, outer cage consists of bars of:

(a) High resistance metal (b) Low resistance

(c) Any one of the above (d) None of these

Ans: **(a)**

Q. 82. A double squirrel cage induction motor possesses:

(a) Good efficiency under running condition

(b) High starting torque

(c) Good speed regulation under normal operating conditions

(d) All of the above

Ans: **(d)**

Q. 83. In double squirrel cage induction motors, the purpose of using high resistance metal bars is to improve:

(a) Starting torque   (b) Starting power factor

(c) Efficiency   (d) Voltage regulation

Ans: **(a)**

Q. 84. A double squirrel cage induction motor is basically a:

    (a) Commutator type motor

    (b) Squirrel cage type motor

    (c) Slip ring type motor

    (d) Conduction type motor

Ans: **(b)**

Q. 85. In double cage motor, the value of upper cage resistance is nearly ...... that of inner cage:

    (a) 2 to 3 times   (b) 5 to 7 times

    (c) Equal to   (d) 4 to 5 times

Ans: **(d)**

Q. 86. In double-squirrel cage induction motor, outer cage possesses:

    (a) High resistance and low leakage reactance

    (b) Low resistance and high reactance

    (c) High resistance and high reactance

    (d) None of these

Ans: **(a)**

Q. 87. In an equivalent circuit diagram of induction motor, two cages (having resistance and reactance) must be connected in:

    (a) Parallel combination

    (b) Either in parallel or series, depending upon load

    (c) Series combination

    (d) None of these

Ans: **(a)**

Q. 88. In induction motors, the term 'Jogging' means:

    (a) Synchronization process

    (b) Cogging phenomenon

    (c) Energizing a motor once or repeatedely to have small movements for mechanisms

    (d) None of the above

Ans: **(c)**

Q. 89. In a particular condition, 3-phase induction motor is running with balanced supply. Suddenly one of the fuses blows off but motor is continuously running. Such type of operation may be called as:

(a) Hunting phenomenon

(b) Single phasing

(c) Crawling phenomenon

(d) Cogging phenomenon

Ans: **(b)**

Q. 90. Suddenly one of the 3-phases of supply to an induction motor fails (may be due to any reason), then the motor is running normally. Now the motor will:

(a) Continue running at the same speed if it was running on light load

(b) Stop and carry heavy current causing permanent damage to the windings, if it was operating over-loaded

(c) Both (a) and (b)

(d) Continue running at half the previous speed

Ans: **(c)**

Q. 91. If the 3-phase balanced supply is given to the 3-phase induction motor and after some time, one phase gets disconnected suddenly, then the motor will:

(a) Run at very high speed

(b) Run at very low speed

(c) Not run at all

(d) Is likely to burn out

Ans: **(d)**

Q. 92. In a 3-phase SRIM, 3-phase supply is provided to rotor and stator winding is short circuited. In this situation:

(a) The rotor would run against the direction of rotating field

(b) The rotor would run at 1/7th N_s

(c) The rotor would run at $N_s/5$

(d) None of these

Ans: (a)

Q. 93. The stator of a 3-phase squirrel cage induction motor is rewound for 6 poles, without any alteration in the rotor. Now the machine will rotate at:

(a) < 1000 rpm

(b) zero rpm

(c) 1500 rpm

(d) None of the above

Ans: (a)

Q. 94. All those 3-phase induction motors, which have open type of slots have:

(a) More starting torque

(b) More breakdown torque

(c) Poor power factor

(d) All of the above

Ans: (d)

Q. 95. The main purpose of using totally enclosed slots or semiclosed slots in 3-phase induction motors is to:

(a) Reduce magnetizing current

(b) Improve power factor and efficiency

(c) Increase efficiency

(d) Both (a) and (c)

Ans: (a)

Q. 96. Two 3-phase induction motors are marked as motor A and motor B. Motor A has deeper and narrow slots and motor B has shallow and wider slots. Which of the following statements is correct?

(a) Motor A, as compared to B, has less starting torque

(b) Motor A has more pull out torque

(c) Motor A has poor operating power factor

(d) All of the above

Ans: (a)

Q. 97. Two, 3-phase squirrel cage induction motors are identical in every respect except that slot depths are different. Machine A has more slot depth compared to machine B. Also both machines have same slot areas. Which of the following statements is correct?

(a) Machine B, as compared to machine A, will have better power factor

(b) Machine B will have less pull out torque and good power factor

(c) Machine B, as compared to machine A, will have more pull out torque and poor power factor

(d) None of the above

Ans: **(c)**

Q. 98. In a 3-phase machine, 4 poles and 48 slots are being used. The number of slots per pole per phase will be:

(a) 2

(b) 8

(c) 4

(d) 1

Ans: **(c)**

Q. 99. The speed control of 3-phase induction motors by supply voltage variation is not done because:

(a) It reduces pull out torque

(b) It increases pull out torque

(c) It reduces power factor of system

(d) None of the above

Ans: **(a)**

Q. 100. Smooth controlling of speed may be obtained by:

(a) Rotor slip power control

(b) Variation of supply frequency

(c) Variation of supply current only

(d) All of the above except (c)

Ans: **(d)**

Q. 101. Which of the following methods will not be used for controlling the speed of squirrel cage induction motor?

(a) Frequency control method

(b) Pole changing process

(c) Line voltage control method

(d) Rotor resistance control method

Ans: **(d)**

Q. 102. Slip changing method of speed control may be used for:

(a) Double squirrel cage and slip ring motors

(b) Squirrel cage and slip ring motors

(c) Slip ring induction motors only

(d) Squirrel cage induction motors only

Ans: **(c)**

Q. 103. In cascade arrangement for speed control of induction motors, the ratio between mechanical power developed and electrical power in rotor of main induction motor having P_1 number of poles will be:

(a) $P_1:P_2$

(b) $P_1 + P_2 = P_2$

(c) $P_2 = (P_1 + P_2)/2$

(d) $P_2:P_1 + 8$

Ans: **(a)**

Q. 104. At subsynchronous speeds, in Kramer system, the electrical power given to the auxiliary commutator machine at slip frequency will be:

(a) Converted into mechanical power and given to the driven shaft

(b) Dissipated as heat loss

(c) Converted to electrical energy

(d) None of the above

Ans: **(a)**

Q. 105. To control, the speed of a 3-phase induction motor during deceleration, ... braking will be employed:

(a) Plugging
(b) DC dynamic braking
(c) Regenerative
(d) None of the above

Ans: **(b)**

Q. 106. Plugging (type of braking, employed for 3-phase induction motors) occurs, when:

(a) Any two supply terminals are open circuited

(b) Any one supply terminal is taken as reference

(c) The supply terminals of any two stator phases are interchanged

(d) Supply voltage is reduced to 1/4th of normal voltage

Ans: **(c)**

Q. 107. Regenerative braking occurs in 3-phase induction motors, when:

(a) Motor speed increases above 3/7th N_s

(b) The number of poles are increased

(c) The load is lowered by a hoisting machine

(d) All of the above

Ans: **(c)**

Q. 108. In dynamic braking of induction motor:

(a) The stator terminals are disconnected and then connected to a DC source

(b) Any two stator terminals are open circuited

(c) DC source is connected to rotor winding

(d) Both (b) and (c)

Ans: **(a)**

Q. 109. An induction motor runs as an induction generator, when its speed is increased beyond:

(a) $N_s/5$
(b) $2N_s$
(c) N_s
(d) $5 N_s$

Ans: **(c)**

Q. 110. Induction generators gives power at:

(a) Leading power factor

(b) Both leading as well as lagging power factors

(c) Lagging power factor

(d) None of these

Ans: **(a)**

Q. 111. To keep the generated voltage frequency constant, with increase in load, in a self-excited induction generator, the speed of induction machine must be:

(a) Increased

(b) Maintained constant

(c) Maintained at N_s only

(d) None of the above

Ans: **(a)**

Q. 112. Armature short circuits can be detected and identified by using:

(a) Growler test   (b) Bar to bar test

(c) Voltage drop test   (d) All of the above

Ans: **(d)**

Q. 113. The bearings of induction motors are designed to bear temperature up to (at normal operating condition):

(a) 192° C   (b) 135° C

(c) 100° C   (d) 95° C

Ans: **(d)**

Q. 114. For the purpose of improving power factor in a 3-phase induction motor, capacitor bank must be used in delta-mode so as to have:

(a) Smaller value of capacitance

(b) Simple capacitance calculation

(c) Good efficiency

(d) Both (b) and (c)

Ans: **(a)**

Q. 115. A single-phase induction regulator has:

 (a) Stator and rotor connected through flexible leads

 (b) A common winding arrangement

 (c) Only rotor winding

 (d) None of these

Ans: **(a)**

Q. 116. To have a voltage regulation of 20%, (i.e. ± 10%), the induction regulator of rating .... of kVA rating of the circuit is needed:

 (a) 10%   (b) 5%

 (c) 7%   (d) 15%

Ans: **(a)**

Q. 117. The primary winding is placed in slots in the surface of the laminated .... core for induction regulator:

 (a) Horizontal   (b) Cylindrical

 (c) Vertical   (d) None of the above

Ans: **(b)**

Q. 118. In 3-phase induction voltage regulators, the phase displacement angle θ may be given as:
(where V_1 = supply voltage and V_r = induced voltage in secondary)

 (a) $\theta = \tan^{-1} [V_r \sin \theta/(V_1 + (V_r \cos \theta)]$

 (b) $\theta = \tan^{-1} [V_r \cos \theta/V_1 - (V_r \sin \theta)]$

 (c) $\theta = \tan^{-1} [V_1 \cos \theta/(V_r + \sin \theta)]$

 (d) $\theta = \tan^{-1} \left[\dfrac{(V_r \cos \theta + V_1)}{\sin \theta} \right]$

Ans: **(a)**

Q. 119. Which of the following statements is not the advantage of a 3-phase induction voltage regulator?

 (a) It is simple in construction

 (b) It is reliable in operation

 (c) It gives small magnetizing current

 (d) None of the above

Ans: **(c)**

Q. 120. An induction voltage regulator is rated by:
  (a) The voltage
  (b) The load current passing through the series winding
  (c) Voltage variations in the connected system
  (d) All of the above except (c)
Ans: (d)

MISCELLANEOUS

Q. 1. The purpose of using blades in SCIM is to:
  (a) Balance the rotor assembly
  (b) Reduce the humming noise
  (c) Improve mechanical strength
  (d) Facilitate cooling of rotor
Ans: (d)

Q. 2. What must be the value of spatial displacement between stator and rotor magnetic field, if the power factor of induction motor is equal to 0.866?
  (a) 120°   (b) 45°
  (c) 180°   (d) None of the above
Ans: (a)

Q. 3. In a 3-phase induction motor, the stator winding is short circuited and fed from the rotor side. What will be the frequency of currents flowing in short circuited stator?
  (a) Slip frequency   (b) Supply frequency/2
  (c) Supply frequency/4   (d) 25 Hz
Ans: (a)

Q. 4. A 3-phase, 6 pole induction frequency converter is driven by prime mover. The converter is connected to 60 Hz, 3-phase supply on the primary. What will be the frequencies of possible outputs from the converter, if the speed of prime mover is kept at 3000 rpm?
  (a) 120 Hz, 50 Hz
  (b) 90 Hz, 210 Hz

(c) 50 Hz, 90 Hz

(d) 60 Hz, 50 Hz

Ans: **(b)**

Q. 5. **In a 3-phase squirrel cage induction motor, the value of rotor leakage reactance is nearly two times its resistance at standstill condition. What will be the frequency at which maximum value of torque is obtained at starting?**

(a) 25 Hz   (b) 45 Hz

(c) 2 Hz   (d) 20 Hz

Ans: **(a)**

Q. 6. **What will be the value of slip at maximum torque in a 3-phase induction motor (stator impedance negligible):**

(a) $r_2/2x_2$

(b) $r_2/\sqrt{\left[(r_2)^2+(x_2)^2\right]}$

(c) r_2/x_2

(d) $\sqrt{\left[r_2/x_2\right]}$

Ans: **(c)**

Q. 7. **When the load is increased in a squirrel cage induction motor, then:**

(a) Stator current decreases

(b) Power factor reduces to 0.5

(c) Slip increases

(d) Both (a) and (b)

Ans: **(c)**

Q. 8. **In induction motors, synchronous watt means:**

(a) Rotor output in watts

(b) Mechanical power output in watts

(c) The torque, which under synchronous speed would create a power of 1 watt or power input to rotor in watts

(d) Output torque in watts

Ans: **(c)**

Q. 9. In the equivalent circuit of a 3-phase induction motor, the resistance R_0 of exciting branch represents:

(a) Rotor copper losses

(b) Friction and windage losses

(c) Stator copper losses

(d) Stator core losses

Ans: (d)

Q. 10. In a slip ring induction motor, with the increment in rotor resistance:

(a) Starting torque increases and efficiency decreases

(b) Starting torque decreases and power factor increases

(c) Efficiency increases and power factor decreases

(d) Only power factor of circuit improves

Ans: (a)

Q. 11. The main purpose of using double cage rotors is to:

(a) Reduce rotor copper losses only

(b) Reduce starting current only

(c) Increase starting torque

(d) Increase break down torque

Ans: (c)

Q. 12. For which type of 3-phase induction motors, direct-on-line starting is employed?

(a) Only wound rotor motor

(b) Only double squirrel cage induction motor

(c) Both wound rotor and squirrel cage motors of small power capacity

(d) Both wound rotor and squirrel cage motors of large power capacity

Ans: (c) .

Q. 13. Star-delta starter is same as autotransformer (starter) with tapping at:

(a) 58%

(b) 60%

(c) 33%

(d) 45%

Ans: **(a)**

Q. 14. Irrespective of supply frequency, the torque developed by a squirrel cage induction motor is same whenever ..... is same:

(a) Slip speed   (b) Supply frequency

(c) Induced emf   (d) None of the above

Ans: **(a)**

Q. 15. At which load, the power factor of a squirrel cage induction motor is low?

(a) Rated load   (b) 1/2 rated load

(c) 3/4th rated load   (d) Light load

Ans: **(d)**

Q. 16. For 3-phase induction motors, torque developed depends upon:

(a) Rotor emf and synchronous speed

(b) Total number of poles and stator current

(c) Rotor current, number of poles and synchronous speed

(d) Rotor emf, rotor power factor and rotor current

Ans: **(d)**

Q. 17. What is the ratio of starting to full load torque, if $I_{sc} = 6$, I_f and $S_f = 5\%$

(a) 1.2   (b) 1.6

(c) 1.8   (d) 1.75

Ans: **(c)**

Q. 18. A 3-phase induction motor is switched on with AC supply but suddenly one of the phases gets disconnected. In this condition:

(a) Motor is likely to burn out quickly unless immediately disconnected

(b) Heavy current will flow through motor windings

(c) Motor will run slowly

(d) Motor refuses to run after attaining zero speed

Ans: (a)

Q. 19. Two, 3-phase induction motors having 6 and 5 poles are connected in cascade arrangement with 60 Hz, 3-phase supply source. The values of different speeds are:

(a) 720, 1200, 1500 and 3600

(b) 720, 1500, 1250 and 500

(c) 1500, 750, 500 and 3000

(d) 1500, 1200, 720 and 3600

Ans: (a)

Q. 20. So as to obtain higher value of starting torque in a 3-phase slip ring induction motor:

(a) Additional resistance must be connected across the slip ring terminals

(b) Windings of rotor side must be permanently connected in star mode

(c) Supply voltage must not be more than 270 volts

(d) None of these

Ans: (a)

Q. 21. Sometimes SCIM runs at a very low speed. This pheno-menon is known as:

(a) Hunting   (b) Skewing

(c) Crawling   (d) Cogging

Ans: (c)

Q. 22. A 3-phase induction motor needs a starter, because:

(a) The induced emf in secondary winding is high during starting instant

(b) The value of currents in the windings are very high at starting

(c) It provides sufficient starting torque

(d) Both (a) and (b)

Ans: (d)

Q. 23. A 3-phase induction generator is mostly used at:

(a) Nuclear power plant

(b) Hydro power plant

(c) Thermal power station

(d) Wind power station

Ans: **(d)**

Q. 24. The data for a 3-phase induction motor is as follows:
Rating——5 hp, 400 V, 4 Pole, 50 Hz
Stator slots-36, Rotor slots-40
In this condition, the rotor may crawl synchronously at:

(a) 150 rpm   (b) 780 rpm

(c) 500 rpm   (d) 550 rpm

Ans: **(a)**

Q. 25. In induction motors, the main consideration is:
Number of stator slots–Number of rotor slots is not equal to P, $2P$ or $5P$
Now it becomes very important so as to avoid:

(a) Cogging   (b) Hunting

(c) Synchronous cusps   (d) Crawling

Ans: **(c)**

Q. 26. Whenever a 3-phase, 350 V, 50 Hz, 1440 rpm induction motor is switched to 420 V, 60 Hz supply source, then the torque:

(a) Will increase

(b) Remains unaltered

(c) Nothing can be said about the concerned matter

(d) None of the above

Ans: **(b)**

Q. 27. In the "injected emf in the rotor circuit" method of speed control of induction motors, the injected emf must be of:

(a) Slip frequency (S_f)

(b) 2 × supply frequency

(c) Supply frequency

(d) $(2 - S)$ × Supply frequency

Ans: **(a)**

Q. 28. **A 3-phase induction motor is allowed to run at super synchronous speed. For self-excitation, the machine:**

(a) Draws reactive power from the mains

(b) Gives reactive power to mains

(c) Draws apparent power from the mains

(d) None of the above

Ans: **(a)**

Q. 29. **In the circle diagram of induction motor, the output line is the line joining the tip of:**

(a) The no load current phasor to the point corresponding to slip = 1

(b) The no load phasor (current) to the point corresponding to slip = ∞

(c) Cannot be represented in circle diagram

(d) None of the above

Ans: **(a)**

Q. 30. **Skewing phenomenon reduces:**

(a) Parasitic torque only

(b) Pull out torque only

(c) Starting and pull out torque, parasitic torque, noise

(d) Pull out torque and increases starting torque

Ans: **(c)**

Q. 31. **In an induction motor, V/f ratio is kept constant and frequency of voltage applied to the motor is increased. In this case, the maximum motor torque will:**

(a) Reduce

(b) Remain unaltered

(c) Remain constant but will occur at smaller values of slip

(d) Reduce and will occur at smaller values of slip

Ans: **(c)**

Q. 32. Skewing phenomenon reduces torque:

(a) Due to rotating fields

(b) Due to 3rd harmonics only

(c) Due to slot harmonics

(d) Due to 5th harmonics only

Ans: **(c)**

Q. 33. A 3-phase, slip ring induction motor is allowed to operate with slip energy recovery in the constant torque mode, when it delievers an output power (P_0) at slip S. What must be the value of maximum power available for recovery at the rotor terminals?

(a) $P_0 S$ (b) $P_0 / 2 - S$

(c) $P_0 / 2 + S$ (d) P_0 / S

Ans: **(a)**

Q. 34. A 3-phase induction motor is running at slip S. What must be the value of efficiency (approximately)?

(a) $1 + S/1 - S$ (b) $1 - S/1 + S$

(c) S (d) $S + 1/S$

Ans: **(b)**

Q. 35. When a 6 pole, 50 Hz slip ring induction motor is supplied at the rated voltage and frequency, with slip ring open circuited, developes approximately 100 volts between any two slip rings. In a particular situation, if the rotor is rotated at 1000 rpm by external means (opposite to stator field), what will be the frequency of voltage across slip rings?

(a) Zero   (b) 25 Hz

(c) 2 Hz   (d) 50 Hz

Ans: **(a)**

Q. 36. In 3-phase systems, a direct-on-line starter is used for starting motors:

(a) Above 5 hp   (b) Up to 5 hp

(c) Above 10 hp   (d) Up to 10 hp

Ans: **(b)**

Q. 37. The full load and maximum efficiency of a 3-phase SCIM are 0.8 and 0.9 respectively. It is allowed to run at a slip of 0.6 (reduced voltage starting). At this operating point, motor-efficiency will be:

(a) Less than 0.4

(b) Less than 0.6

(c) Greater than 0.4

(d) Greater than 0.6

Ans: (a)

Q. 38. At normal voltage, the value of slip is 2%. What will be the value of slip, when the same amount of torque is developed at 10% below normal voltage?

(a) 2.47%

(b) 3.47%

(c) 3.25%

(d) 3.33%

Ans: (a)

Q. 39. In a 3-phase SRIM, input voltage is 440 volts and voltage at slip rings is 110 volts. Whenever an input of 110 volts is given to the slip rings, what will be the stator voltage?

(a) Approximately 440 volts

(b) Approximately 120 volts

(c) Approximately 55 volts

(d) Approximately 27.5 volts

Ans: (a)

Q. 40. In induction motors, the number of stator poles may be varied by:

(a) Pole amplitude modulation

(b) Method of consequent poles

(c) Multiple stator windings

(d) All of the above

Ans: (d)

FILL IN THE BLANKS

1. Polyphase induction motor is a …….. excited machine.

Ans: Singly

2. An asynchronous machine may be considered as a rotating ……..

Ans: Transformer

3. The air gap length is kept …….. in induction motors.

Ans: Small

4. …….. induction motor is most widely used in industries.

Ans: Squirrel cage

5. For smaller machines, the value of standstill open circuit slip ring voltage lies in the range of …….. volts.

Ans: 100 to 400

6. …….. speed is the speed, at which the field produced by primary currents will revolve.

Ans: Synchronous

7. Slip frequency is the product of slip (S) and …….. .

Ans: Supply frequency (f)

8. Maximum value of torque varies inversely as …….. of the rotor.

Ans: Standstill reactance

9. Maximum running torque is …….. of rotor resistance in three-phase induction motor.

Ans: Independent

10. In three-phase induction motors, maximum torque varies directly as the square of ……..

Ans: Supply voltage

11. At standstill condition, the motor is just equivalent to a three-phase …….. with …….. winding short circuited.

Ans: Transformer; Secondary

12. …….. motors are employed in cranes and elevators for acceleration of heavy loads.

Ans: Wound rotor

13. In three-phase induction motors, whenever 3-phase supply is given to the stator terminals, the resultant flux is 1.5 times the ....

Ans: **Maximum flux**

14. Three-phase induction motors have good .... load capacity.

Ans: **Over**

15. Stator is built up of high grade alloy .... steel laminations.

Ans: **Silicon**

16. Cage rotor consists of a .... core with slots nearly parallel to shaft-axis.

Ans: **Cylindrical laminated**

17. In three-phase induction motors, the direction of rotation of resultant flux in the air gap depends upon ....

Ans: **Phase sequence**

18. Slip speed expresses the rotor speed relative to ....

Ans: **Field**

19. For large rating motors, full load slip varies from .... percent.

Ans: **2 to 5**

20. Three-phase induction motor runs at nearly .... speed.

Ans: **Constant**

21. The purpose of laminating the stator core is to reduce .... losses.

Ans: **Eddy current**

22. In large induction motors, the air gap is usually from 1.0 mm to .... mm.

Ans: **1.5**

23. Total number of slip rings used in squirrel cage induction motors are ....

Ans: **Zero**

24. The speed of induction motor is always .... compared to synchronous speed.

Ans: **Lesser**

25. At starting moment, the value of slip for induction motor will be ....

Ans: 1 (one)

26. In induction motors, rotor current frequency is given as the product of per unit slip and ....

Ans: Supply frequency

27. The maximum torque in induction motors may also be known as........ torque.

Ans: Break down or pull out

28. In induction motors, torque will be maximum, when resistance of secondary winding will be slip times ....

Ans: Rotor leakage reactance

29. Starting torque will be proportional to square of ....

Ans: Applied voltage

30. In standard squirrel cage induction motors, the break down torque lies between ....% to .... % of full load torque.

Ans: 200; 300

31. Rotor and stator copper losses vary with the .... of their respective currents.

Ans: Square

32. The stray load losses are nearly .... percent of power supplied to motor.

Ans: 0.5

33. The difference between gross torque (T_g) and .... torque is equal to torque lost due to friction and windage losses in the motor.

Ans: Shaft

34. Induction motor is a .... transformer, with secondary (rotor) winding short circuited.

Ans: Static

35. In induction motors, induced torque may also be called as .... torque.

Ans: Electromagnetic

36. Standstill torque may also be known as …….. torque.

Ans: Starting

37. Pull out torque in induction motors may also be called as …….. torque.

Ans: Break down

38. For a typical induction motor, starting torque is approximately …….. times the rated full load torque.

Ans: 1.5

39. Heat run test is performed to determine the …….. rise of various parts of motor, while it is running at rated speed.

Ans: Temperature

40. At no load condition, the power input to 3-phase induction motor will be equal to the addition of core loss, stator copper loss and ……. losses.

Ans: Windage and friction

41. At no load condition, the power factor would lie in the range of …….. to ……..

Ans: 0.05; 0.15

42. Blocked rotor test may also be known as …….. test performed on 3-phase induction motors.

Ans: Short circuit

43. The main purpose of heat run test is to determine the actual maximum …….. attained, when the machine is continuously running under loaded conditions.

Ans: Temperature

44. In induction motors, developed mechanical power is the difference between rotor input and …….. loss.

Ans: Rotor copper

45. The purpose of using starter is to reduce the heavy …….. current.

Ans: Starting

46. Small sized motors up to …….. kW rating may be started by direct-on-line starters.

Ans: 5

47. In induction motors, variable losses include …….. and …….. losses.

Ans: **Stator; Rotor copper**

48. Induction motor always operates at …….. power factor.

Ans: **Lagging**

49. A star-delta starter is used for a cage-motor designed to run normally on …….. connected …….. winding.

Ans: **Delta; Stator**

50. In the star-delta starting, starting torque reduces to nearly …….. of the torque obtained from direct-on-line starting.

Ans: **One third**

51. For both star and delta connected motors, an …….. is suitable for starting purpose.

Ans: **Autotransformer starter**

52. Star-delta starter is cheaper compared to an …….. starter.

Ans: **Autotransformer**

53. Slip ring induction motors may be started with …….. line voltage.

Ans: **Full**

54. Whenever, 3-phase supply is given to the 3-phase stator winding, a …….. field will be created in the airgap of induction motor.

Ans: **Rotating magnetic**

55. No load test on a 3-phase induction motor is equivalent to …….. on a transformer.

Ans: **Open circuit test**

56. Locked rotor test must be performed at a ……… frequency.

Ans: **Reduced**

57. Cogging phenomenon is also called as ………

Ans: **Magnetic or teeth locking**

58. Cogging may be reduced by using …….. rotor.

Ans: **Skewed**

59. The rotor of an induction motor fails to start, if the number of stator and rotor slots are either .... or have an integral ratio.

Ans: Equal

60. Number of stator and .... slots are not made to be ....

Ans: Rotor; Equal

61. In induction motors, crawling phenomenon may be reduced by reducing .... and .... harmonics.

Ans: Fifth; Seventh

62. Whenever primemover of an induction machine rotates the rotor at a speed greater than .... speed, the value of .... becomes negative.

Ans: Synchronous; Slip

63. No .... current excitation is required by induction generator.

Ans: Direct

64. An induction generator may also be called as .... generator.

Ans: Asynchronous

65. The maximum possible induced torque in the generating mode (as according to torque-slip characteristic) is called as .... torque of generator.

Ans: Pushover

66. Induction generator has a small size per kilowatt .... power.

Ans: Output

67. Induction generator may be called as .... generator, because the rotor speed is different from the .... speed.

Ans: Asynchronous; Synchronous

68. In induction motors, the effect of crawling may be minimized by skewing, chording and .... winding.

Ans: Integral slot

69. Induction voltage regulator acts as a .... transformer.

Ans: Step down

70. When the induction voltage regulator is not provided with short circuited …….. winding, it will not give the satisfactory results.

Ans: Compensating

71. The compensating winding tends to keep I^2R losses …….. for a given load current.

Ans: Constant

72. No …….. or …….. winding is needed in 3-phase induction voltage regulator.

Ans: Compensating; Tertiary

73. Induction voltage regulator is more …….. compared to transformer with tap changing mechanism.

Ans: Costly

74. 3-phase IVRs may also be known as …….. transformers.

Ans: Phase shifting

75. When the voltage applied to stator winding is reduced then …….. current will also be reduced accordingly.

Ans: Short circuit

76. Full voltage starting may also be called as …….. starting.

Ans: Direct-on-line

77. If the induction motor is directly given AC supply, then the current at starting will be …….. current.

Ans: Short circuit

78. A manual autotransformer starter is actually a …….. switch.

Ans: Multi pole double throw

79. The main drawback of autotransformer starter includes …….. power factor and …….. cost in lower output rating motors.

Ans: Low; Higher

80. Star delta starters are used for starting 3-phase SCIMs of rating between …….. and …….. kW.

Ans: 4; 20

81. The speed control of induction motor may be done from rotor side by varying the resistance in the ....

Ans: Rotor circuit

82. Variation of rotor-resistance for controlling the speed of induction motor is not convenient (for controlling speed) at .... torque.

Ans: constant

83. Various motor speeds may be obtained by .... connections of induction motors.

Ans: Cascade

84. Kramer's system of speed control is actually based on the extraction of power from the .... circuit of a wound rotor induction motor via ....

Ans: Rotor; Slip rings

85. Scherbius drive of speed control is .... but superior to Kramer's drive, when good speed control is needed near synchronous speed.

Ans: Costlier

86. Plugging is also known as .... braking.

Ans: Counter current

87. During plugging period, the emf induced in the rotor is ....

Ans: Very high

88. .... is mainly useful for reversing drives where braking and starting up of induction motor in reverse direction comprises stages of the same continuous process.

Ans: Plugging

89. .... braking may be obtained by disconnecting the stator winding from AC supply and exciting it from a DC source to produce a .... DC field.

Ans: Dynamic or Rheostatic; Stationary

90. The shapes of rotor teeths and slots affect the ...., which affects starting current and .... torque.

Ans: Reactance; Maximum

91. .... slot windings are not used for induction motor stators.

Ans: Fractional

92. Cogging and synchronous cusps may be reduced by ....

Ans: Skewing

93. The double squirrel cage induction motors are designed so as to provide a high .... torque with a .... starting current.

Ans: Starting; Low

94. The pull out torque of a double squirrel cage motor is .... compared to a plain squirrel cage motor.

Ans: Smaller

95. Double cage motor is approximately .... percent more costly compared to plain squirrel cage motor.

Ans: 20 to 30

96. Squirrel cage induction motor has .... rotor overhang .... flux.

Ans: Low; Leakage

97. Slip ring motors possess high value of .... torque with low .... current.

Ans: Starting; Starting

98. A wide range of torque-slip characteristics may be obtained with .... cage motors.

Ans: Double

TRUE / FALSE

1. Three-phase induction motors are mainly employed for industrial applications.

Ans: True

2. An induction machine may also be used as a synchronous generator, if the rotor is rotated by means of prime mover.

Ans: False

3. In induction motors, stator core is supported by windings.

Ans: False

4. Whenever 3-phase supply is given to stator windings, a rotating magnetic field of variable magnitude is produced.

Ans: False

5. In induction motors, if the sequence of supply to stator winding terminals is changed, the direction of rotation of the stator magnetic field is reversed.

Ans: True

6. Three-phase induction motors run at a speed slightly lower than a predetermined speed (obtained by load circuit).

Ans: False

7. In squirrel cage induction motors, additional resistance may be connected in the rotor circuit to control speed.

Ans: False

8. In induction motors, the resultant air gap flux rotates with synchronous speed.

Ans: True

9. The stator of a 3-phase induction motor is similar to that of a synchronous machine.

Ans: True

10. Slip ring motors are never employed, where greater starting torque is required.

Ans: False

11. In induction motors, the no load current changes from nearly 30% to 50 % of full load current.

Ans: True

12. The frequency of rotor emf is same as the supply frequency at all running conditions.

Ans: False

13. The rotor of an induction motor can never attain synchronous speed.

Ans: True

14. The relative speed between rotating magnetic field and rotor conductors is zero at standstill condition.

Ans: **False**

15. The air gap power is the difference of stator power input and stator losses.

Ans: **True**

16. Stray load loss occurs in iron as well as in conductors.

Ans: **True**

17. In induction machines, fixed losses is the difference of core losses and friction losses.

Ans: **False**

18. Whenever 3-phase voltage is given to stator terminals of a 3-phase induction motor, a rotating air gap flux Ø is established.

Ans: **True**

19. The exciting current changes from 10% to 15% of full load current in induction motors.

Ans: **False**

20. In induction motors, maximum internal torque may also be called as pull out torque or break down torque.

Ans: **True**

21. In a wound rotor induction motor, the starting torque is independent of resistance.

Ans: **False**

22. The value of no load slip is very small in induction motors.

Ans: **True**

23. In induction motors, the no load stator current is nearly 70% of rated current.

Ans: **False**

24. In polyphase induction motors, no load test can also be referred to as running light test.

Ans: **True**

25. While performing blocked rotor test, the entire input power is wasted in stator and rotor windings as I^2R loss.

Ans: **True**

26. At standstill condition, rotor core loss is appreciable.

Ans: **True**

27. The circle diagram of an induction motor may be drawn by using the data obtained from no load and stator resistance test only.

Ans: **False**

28. For three-phase induction motor, the no load power factor varies from 0.1 to 0.3 lagg.

Ans: **True**

29. When full applied voltage is used for 3-phase induction motors at starting instant, then very high starting current is obtained.

Ans: **True**

30. Direct-on-line starting means across the line starting.

Ans: **True**

31. In the stator resistor (or reactor), starting a resistor or a reactor is used in between motor-terminals and the supply terminals.

Ans: **True**

32. While using autotransformers as starters, starting torque is X^2 times, corresponding values with direct-on-line starting.

Ans: **True**

33. Star-delta starter is costly as compared to autotransformer starter.

Ans: **False**

34. Star-delta starter is not used for line voltages exceeding 1.1 kV.

Ans: **False**

35. In wound rotor motors, the rotor winding terminals are connected to three slip rings, mounted on the shaft.

Ans: **True**

36. In an induction generator, the slip S is negative.

Ans: True

37. An induction generator is also called as asynchronous generator.

Ans: True

38. An induction generator requires greater maintenance compared to induction motor and other equipments.

Ans: False

39. Reactive volt-amperes can not be generated by induction generator.

Ans: True

40. Squirrel cage induction motor runs stably at nearly 1/7th of synchronous speed and the phenomenon is called as cogging.

Ans: False

41. Cogging of squirrel cage motors may be easily reduced or eliminated by making the number of rotor slots prime to the number of stator slots.

Ans: True

42. In double cage motors, inner cage has high resistance copper bars.

Ans: False

43. Double cage motor consists of two cages namely outer cage and inner cage.

Ans: True

44. At full load condition, the speed regulation of a 3-phase induction motor is less than 2% (always).

Ans: False

45. Pole changing method for speed control of 3-phase induction motor is used for elevator motors.

Ans: True

46. Rotor rheostat control method is applicable for squirrel cage induction motors.

Ans: False

47. Cascade operation (speed control purpose) may also be known as tandem operation.

Ans: True

48. Kramer's system of speed control may easily be used for motors up to 5 kW rating.

Ans: False

49. Scherbius system is a method, employed for 3-phase induction motors for starting purpose.

Ans: False

50. In the cascade sets, the slip power of main induction motor is fed to an auxiliary induction motor.

Ans: True

51. The Leblanc exciter is a frequency converter.

Ans: True

52. Regenerative method of braking may be used in hoisting type of mechanism.

Ans: True

53. Squirrel cage motors are useful in obtaining nearly constant speed and high overload capacity.

Ans: True

54. In wound rotor induction motor, high value of power factor may be obtained at lightly loaded conditions.

Ans: False

55. Induction motors may operate at lagging and leading power factor both.

Ans: False

56. A squirrel cage induction motor is a constant speed motor.

Ans: True

57. The air gap is one of the main sources of the low power factor at which the induction motor operates.

Ans: True

58. The induction harmonic torque may be reduced by making a proper choice of coil span and by skewing rotor slots.

Ans: **True**

59. The closed type slots have very low leakage reactance.

Ans: **False**

60. For induction motor stators, fractional slot windings are employed.

Ans: **False**

61. Double cage rotor motor is basically a high torque and low starting current motor.

Ans: **True**

62. Double cage motor is nearly 70% costly compared to squirrel cage motor.

Ans: **False**

63. When the number of stator slots is equal to rotor slots, the motor may refuse to start. This is called cogging phenomenon.

Ans: **True**

64. In single-phase induction voltage regulator, secondary winding is connected across the 1-φ supply.

Ans: **False**

65. In induction regulator, when the axes of primary and secondary windings are 90° apart, maximum value of emf is induced in primary.

Ans: **False**

66. When the regulator position is adjusted for any intermediate position, a certain portion of secondary mmf reacts with primary.

Ans: **True**

67. The speed of a three-phase induction motor can be controlled by using frequency control method only.

Ans: **False**

68. Wound rotor motors are preferred, where low starting torque is needed.

Ans: **False**

69. Air gap between the stator and rotor must be kept free from any accumulation of dirt.

Ans: True

70. Locked rotor test at a suitable voltage may be known as routine test.

Ans: True

71. Temperature rise test and high voltage test of a 3-phase induction motor may also be known as type test.

Ans: True

72. An induction regulator provides voltage control in steps.

Ans: False

73. In single-phase induction regulator, stator houses the secondary winding.

Ans: True

74. In three-phase induction regulator, both the stator and rotor houses three-phase windings.

Ans: True

75. For the proper foundation of an induction motor, foundation block must be at least 20 cm longer and 20 cm wider than the motor feet, side rails or bed plates.

Ans: True

VIVA VOCE QUESTIONS

Q. 1. What do you mean by induction motor?

Ans: The motor in which the current is actually induced in the rotor conductors due to the relative motion between rotor conductors and rotating magnetic field, produced by stator current, is called as induction motor.

Q. 2. Induction motors are also called as asynchronous motors, why?

Ans: In these motors, rotor does not turn in synchronism with the rotating magnetic field developed by stator currents, so these motors are also known as "asynchronous motors".

Q. 3. Why the stator and rotor cores of an induction motor are made up of laminated sheets?

Ans: These are made up of laminated sheets, so as to reduce the eddy current losses in the cores.

Q. 4. What is the function of frame, in induction motor?

Ans: The functions are:
(a) To protect the inner part of machine
(b) To support the winding and stator core, etc.

Q. 5. What will happen if the frame of induction motor is not made rigid?

Ans: If the frame is not rigid, then the rotor of induction motor will not remain concentric with stator, giving rise to unbalanced magnetic pull.

Q. 6. Why the stator core is assembled of high grade low electrical loss, silicon steel punchings?

Ans: In order to reduce eddy current and hysteresis losses in the core.

Q. 7. What are the two main types of rotor in an induction motor?

Ans: The two types are:
(a) Squirrel cage rotor   (b) Wound rotor

Q. 8. Why the external resistance can not be added in the rotor circuit in an SCIM?

Ans: Because the rotor winding is permanently short circuited in squirrel cage construction.

Q. 9. Which type of slots are preferred in induction motors?

Ans: The slots on the rotor are either of semi-closed type or of totally closed type. Both these slots may be used.

Q. 10. Which rotor is used in squirrel cage induction motor?

Ans: Squirrel cage type

Q. 11. Which rotor is used in slip ring induction motor?

Ans: Wound rotor

Q. 12. What is the meaning of synchronous speed?

Ans: The speed at which the field produced by the primary currents will revolve is called the synchronous speed of the motor. It is expressed as $N_s = 120f/P$, where f is applied frequency and P is the number of poles on the stator.

Q. 13. In induction motors, the stator windings are almost always short-pitched, why?

Ans: Actually it is short pitched because it reduces copper weight and winding resistance. It also reduces the values of leakage reactance and harmonic torque disturbances.

Q. 14. Explain the principle of operation of induction motor?

Ans: When the stator windings of a 3-phase induction motor are connected to a 3-phase balanced supply, they receive balanced 3-phase currents, which develop a rotating flux wave. This rotating flux cuts the rotor conductors and induces emf in it. As these conductors are short-circuited, so these induced emfs develop current in the rotor. Finally these currents and the flux wave interact with each other so as to produce the torque in the rotor. Now according to Lenz's law, this developed torque must oppose the cause to which it is due that is cutting of flux lines by the rotor conductors. So the developed torque allows the rotor to rotate in the direction of the flux wave so as to reduce the relative speed between the flux wave and the conductors (rotor). Thus finally the rotor of induction motor rotates.

Q. 15. On which factor, the direction of rotation of resultant flux (in the air gap) depends?

Ans: It depends upon the phase sequence.

Q. 16. What do you mean by slip speed in an induction motor?

Ans: The actual difference between the synchronous speed and actual rotor speed is called as "slip speed". It is expressed as $N_s - N$, where N_s and N are synchronous speed and actual rotor speed respectively.

Q. 17. What is the meaning of fractional or per unit slip?

Ans: The slip speed expressed as a fraction of the synchronous speed is called the per unit slip or fractional slip.

$S = N_s - N/N_s$ = PU slip, Also % slip = $(N_s - N/N_s) \times 100$

Q. 18. Why an induction motor takes more value of magnetizing current as compared to a transformer of the same rating?

Ans: It is due to the presence of air gap in induction motor.

Q. 19. What are the benefits of a three-phase induction motor?

Ans: The benefits are:
(a) Possesses good efficiency
(b) Maintenance cost is low
(c) Construction is simple etc.

Q. 20. What are the advantages of external resistance in the rotor circuit of a slip ring induction motor?

Ans: It may help in
(a) Increasing starting torque
(b) Reducing the high starting current
(c) Adjusting motor speed

Q. 21. At which condition, the torque developed will reach at maximum position in induction motor ?

Ans: When $R_2 = SX_2$, then maximum torque will be achieved in induction motor.

Q. 22. What is the relation between the full load and maximum torque?

Ans: $T_f/T_{max} = 2aS_f/a^2 + S_f^2$

where a = slip corresponding to maximum torque and S_f = full load slip

Q. 23. What is the relation between starting and maximum torque?

Ans: $T_{st}/T_{max} = 2a/1 + a^2$

Q. 24. What is the meaning of stray load losses in an induction motor ?

Ans: The stray load losses in an induction motor are as follows:

(a) Harmonic losses in rotor conductors

(b) Additional fundamental frequency and high frequency losses in the iron

(c) Rotor iron losses

(d) Circulating current losses in the stator winding

Q. 25. What is the relation between the mechanical power developed and gross torque in an induction motor?

Ans: $P_{mech} = T_g \times (2. \pi.N/60)$ W

Q. 26. What will be the ratio of the following three terms related to induction motor?

(a) Power input to rotor (P_2)

(b) Gross rotor output (P_{mech}) power and

(c) Rotor copper loss (P_{Cu})

Ans: The ratio between above terms will be

$P_2 : P_{mech} : P_{Cu} = 1 : (1 - S) : S$

Q. 27. What do you mean by synchronous watt?

Ans: It is a new unit of torque and defined as the torque which developes a power of 1 watt at the synchronous speed of the motor.

Q. 28. On which factor, the magnitude of the starting torque and maximum torque depend?

Ans: Both of them depend upon the rotor resistance.

Q. 29. What are the various tests, performed on a 3-phase induction motor?

Ans: These tests are:

(a) No load test

(b) Blocked rotor test

(c) Stator resistance test

(d) Heat run test

Q. 30. What is the main purpose of performing no-load test on a 3-φ induction motor ?

Ans: The purpose of no-load test is to obtain core losses, friction and windage losses, magnetizing current, power factor at no load condition and the parameters of the magnetizing branch of the equivalent circuit.

Q. 31. Which test is analogous to the short circuit test of a transformer?

Ans: It is the blocked rotor test.

Q. 32. From which test, the value of rotor resistance can be obtained?

Ans: The rotor resistance can be determined from blocked rotor test.

Q. 33. Why the heat run test is also known as temperature rise test?

Ans: Actually heat run test is done so as to obtain the temperature rise of various parts of the motor while it is running at rated speed. That is why, it is also known as temperature rise test.

Q. 34. In induction motor, iron losses in the rotor are neglected, why?

Ans: Actually it is due to very low frequency of emf induced in the rotor.

Q. 35. Which tests are required to draw the circle diagram of a 3-phase induction motor?

Ans: These tests are:
(a) No load test
(b) Blocked rotor test

Q. 36. Which current can be obtained from short circuit test?

Ans: The short circuit current I_{sc} corresponding to normal voltage applied to stator winding is obtained from SC test in induction motor.

Q. 37. What is the advantage of circle diagram of induction motor?

Ans: The advantages of circle diagram are as follows:

    (a) Very fast estimation of the induction motor operating characteristics

    (b) Simplicity

Q. 38. Which data is obtained from locked rotor test of an induction motor?

Ans: The data obtained is:

    (1) Power factor at short circuit condition ($\cos\varphi_{sc}$)

    (2) Short circuit current with normal voltage applied to stator winding

    (3) Total value of resistance and reactance of induction motor as referred to stator side (i.e. the values of R_{01} and X_{01})

Q. 39. What do you mean by torque line in the circle diagram of an induction motor?

Ans: It is the line which seperates the stator and rotor copper losses.

Q. 40. An induction motor always operates at lagging power factors for any value of load, why?

Ans: The reasons are as follows:

    (1) Stator and rotor leakage reactance drops increase the power factor angle between applied voltage and current. Consequently the final effect of leakage reactances is to decrease the power factor.

    (2) The magnetizing current, required for creating magnetic flux, lags the applied voltage by 90°.

Q. 41. What are the advantages and disadvantages of reduced voltage starting?

Ans: Advantage: Starting current is reduced (for safety purpose)

Disadvantage: Objectionable reduction in the starting torque

Q. 42. When the induction motor is directly switched, how much torque will be developed?

Ans: When directly switched, the torque developed will be nearly 1.5 to 2.5 times their full load torque.

Q. 43. How can the starting torque of an induction motor be improved?

Ans: Starting torque can be improved by increasing the resistance of rotor circuit.

Q. 44. On which factors, the choice of any particular method of starting of SCIM depends?

Ans. These factors are:
   (a) The type of driven load
   (b) Motor size
   (c) Design of motor
   (d) Capacity of the power lines

Q. 45. Name the methods of starting for squirrel cage induction motors.

Ans: These starting methods are as follows:
   (a) Auto starter method
   (b) Star-Delta starter method
   (c) Primary resistor or reactor method

Q. 46. Compare the series reactor and series resistor w.r.t. resistance/reactor starting method?

Ans: Actually the series resistor is cheaper compared to series reactor, but this reactor has lower energy loss and is more attractive and beneficial in reducing the voltage.

Q. 47. For which motors, Star-Delta starter is used?

Ans: This method is preferred in those motors, which are built to run normally with a delta connected stator winding.

Q. 48. Which starting method is used for machine tools and motor-generator sets?

Ans: Star-Delta starter

Q. 49. Why the slip ring motors can be started under load?

Ans: Since external resistance is added in the rotor circuit, so a slip ring motor will develop a high starting torque and moderate starting current, hence these motors can be started under load.

Q. 50. What do you mean by crawling phenomenon in an induction motor?

Ans: Sometimes induction motor tries to run at nearly 1/7th of their synchronous speed in a stable manner. This phenomenon is known as crawling phenomenon.

Q. 51. What is cogging phenomenon in an induction motor?

Ans: Sometimes squirrel cage rotor of SCIM refuses to start, (when the applied voltage is low). It actually happens when the number of stator teeth and rotor teeth becomes equal and is due to the magnetic locking between the teeths of stator and rotor. This phenomenon is called as cogging phenomenon.

Q. 52. What are the various speed control methods of induction motors?

Ans: The various methods of speed control are:
(a) From stator side: (1) By changing the applied voltage
   (2) Variation of frequency
   (3) By changing the number of poles
(b) From rotor side: (1) By using the additional injected emf in the rotor circuit
   (2) By varying the resistance in the rotor circuit

Q. 53. Why the speed control method of induction motor by variation of supply voltage has limited use?

Ans: It has limited use because
(1) Speed control range is very limited in the downward direction
(2) A large change in voltage is needed for small change in speed
(3) The voltage can not be raised beyond rated voltage (due to magnetic saturation)

Q. 54. Which of the speed control methods is similar to the series–parallel control of DC motor?

Ans: Concatenation method of speed control.

Q. 55. What is double squirrel cage motor?

Ans: This motor consists of two cages namely outer cage and inner cage. Outer cage has bars of high resistance metal and inner cage consists of low resistance copper bars.

Q. 56. Which one of the two cages developes maximum torque at starting?

Ans: Outer cage. It is because outer cage has high value of resistance and low ratio of reactance to resistance.

Q. 57. What are the various methods of electrical braking of induction motors?

Ans: Various braking methods are as follows:
(1) Plugging (also called as counter-current braking)
(2) Dynamic (Rheostatic) braking
(3) Regenerative braking

Q. 58. How the rheostatic braking is obtained?

Ans: This rheostatic braking can be obtained by disconnecting the stator winding from AC supply and exciting it from a DC source to create a stationary DC field.

Q. 59. Why the regenerative braking method is seldom used?

Ans: In this braking method, the possibility of braking occurs only at supersynchronous speeds. That's why the regenerative method is seldome used.

Q. 60. What do we mean by induction generator?

Ans: When an induction machine is coupled to a prime mover and is driven at a speed greater than synchronous speed, then the machine is called as induction generator.

Q. 61. What is induction voltage regulator?

Ans: It is basically a step-down transformer, by which the secondary voltage may be varied from zero to a certain maximum value. It consists of two windings (primary and secondary winding). The primary winding is connected across the circuit to be regulated and the secondary is connected in series with the circuitry.

Q. 62. What are the types of induction voltage regulator?

Ans: It may be of single-phase or three-phase, so called as single-phase induction regulator and three-phase induction voltage regulator.

Q. 63. In a 3-phase induction voltage regulator, tertiary winding is not needed, why?

Ans: For each position of setting of the regulator, each secondary winding is magnetically coupled to one or more of the primary windings. That's why no tertiary winding is needed.

Q. 64. What are the advantages of induction voltage regulator?

Ans: These advantages include:
   (a) Reliable features
   (b) Can withstand overloads in a well manner
   (c) Simplicity
   (d) Gives stepless voltage variation without any arcing problem

Q. 65. Why the rotor slots are skewed?

Ans: These slots are skewed so as to obtain a uniform torque and reduce the magnetic humming noise. In addition to this, skewing also reduces the magnetic locking of the stator and rotor.

Q. 66. How can the direction of rotation of 3-phase induction motor be reversed?

Ans: It may be reversed by interchanging the connection to the supply of any two leads of the motor.

Q. 67. The efficiency of 3-phase induction motor is less compared to transformer, why?

Ans: The reasons for low efficiency of 3φ induction motor are:

(1) Friction and windage losses in induction motor

(2) Large value of magnetizing current

(3) Presence of air gap

Q. 68. What do you mean by single phasing?

Ans: Single phasing means the operation of motor with one stator terminal accidently disconnected.

Q. 69. What must be the synchronous speed, if the full load speed of an induction motor is nearly 965 rpm?

Ans: It must be 1000 rpm.

Q. 70. What is the relation between the torque and applied voltage in an induction motor?

Ans: The relation is——$T \propto SV^2$.

Q. 71. What must be the frequency of induced emf in the rotor, if the slip and normal frequency are S and f respectively?

Ans: Frequency of induced emf = Sf.

Q. 72. What are the advantages of slip ring induction motor?

Ans: The advantages of slip ring induction motor are:

(a) The speed can be easily controlled by changing external resistance in the rotor circuit

(b) No abnormal heating during starting period

(c) High starting torque

(d) Nearly constant speed output, etc.

Q. 73. What are the advantages of a squirrel cage motor?

Ans: The advantages of cage motor are as follows:

(1) Rugged construction

(2) Better cooling

(3) Low maintenance and initial cost

(4) Good pull out torque

(5) Better efficiency and power factor, etc.

Q. 74. What will be the value of torque developed in induction motor, when the slip is zero?

Ans: Torque will also be zero.

Q. 75. What precaution must be taken, while designing the circuit for a Star-Delta starter?

Ans: Care should be taken that the direction of rotation of the motor must remain same, while the change in connections of the stator windings occur from star to delta.

9

Three-Phase Transformer

Q. 1. As compared to a bank of single-phase transformers, advantage of 3-phase transformer is/are:

(a) Less weight

(b) Occupies less floor space

(c) Overall cost 15% less

(d) All of the above

Ans: **(d)**

Q. 2. The three-phase transformer may be of:

(a) Core type (b) Shell type

(c) Both types (d) None of the above

Ans: **(c)**

Q. 3. Which of the following connections is best suited for 3-phase 4 wire service?

(a) Delta–Delta (b) Star–Delta

(c) Delta–Star (d) Star–Star

Ans: **(c)**

Q. 4. In case of three-phase transformers, Star–Star (or Y–Y) connection is most economical for:

(a) Small, high voltage transformers

(b) Large, low voltage transformers

(c) Small, low voltage transformers

(d) Large, high voltage transformers

Ans: **(a)**

Q. 5. In Star–Star connection, what is the phase shift between the phase voltage and line voltage both on the primary and secondary sides?

(a) 30°   (b) 60°

(c) 75°   (d) 120°

Ans: (a)

Q. 6. Delta–Delta connection (Δ–Δ) is used for:

(a) Large, low voltage transformers

(b) Small, low voltage transformers

(c) Large, high voltage transformers

(d) Small, high voltage transformers

Ans: (a)

Q. 7. As compared to Δ–Δ bank, what will be the capacity of the V--V bank of transformer?

(a) 86.8   (b) 50

(c) 57.7   (d) 66.7

Ans: (c)

Q. 8. In a Y–Δ transformer, voltmeter is connected in secondary side (opening a node of delta), the voltmeter will give:

(a) Zero voltage

(b) Phase voltage in secondary side

(c) Line voltage

(d) None of the above

Ans: (a)

Q. 9. In Δ–Δ connection of transformer, if one of the transformer is removed, then capacity of the system will reduce to:

(a) 66.67%

(b) 57.74%

(c) 50%

(d) 33.33%

Ans: (b)

Q. 10. When transformers are connected in Δ–Z (Delta–Zigzag) and Star–Zigzag mode, then what will be the utilisation factors for transformers?

(a) 0.866 and 0.577   (b) 0.755 and 0.792

(c) 0.866 and 0.866   (d) 0.5 and 0.667

Ans: **(c)**

Q. 11. When transformers are operated in V–V bank mode, then what will be the load in VA per transformer?

(a) $1/\sqrt{5}$ × original load in Delta–Delta bank

(b) $1/\sqrt{3}$ × original load in Delta–Delta bank

(c) 3 × original load in Delta–Delta bank

(d) None of the above

Ans: **(b)**

Q. 12. In open-delta connection of transformers, what is the ratio of $(S_{V-V}/S_{\Delta-\Delta})$ VA ratings?

(a) $1/\sqrt{3}$ (b) $1/\sqrt{7}$

(c) $1/\sqrt{2}$ (d) None of the above

Ans: **(a)**

Q. 13. What will be the symbol for Delta–Zigzag transformer?

(a) DZ0 or DZ6   (b) DZ11 only

(c) DZ6   (d) DZ0 only

Ans: **(a)**

Q. 14. In which transformer, the tertiary winding is used?

(a) Star–Delta   (b) Delta–Delta

(c) Star–Star   (d) Delta–Star

Ans: **(c)**

Q. 15. When a V–V system is converted into a Δ–Δ system, increase in capacity of the system is:

(a) 73.2 %   (b) 86.6%

(c) 27.39%   (d) 50.5%

Ans: **(a)**

Q. 16. Which of the following factors affects choice of transformer connections?

(a) Parallel operation with other transformers

(b) Insulation to ground and voltage stresses

(c) Need for partial capacity with one circuit out of service

(d) All of the above

Ans: **(d)**

Q. 17. In case of a 3-phase transformer, advantages of Delta–Delta transformation include:

(a) When one of the transformer fails, the remaining two transformers will continue to supply 3-phase power

(b) If however, 3rd harmonic component is present, it circulates in the closed path and does not appear in the output voltage waveform

(c) This connection is suitable for both balanced and unbalanced loading conditions

(d) All of the above

Ans: **(d)**

Q. 18. In case of Y–Y connection, problems of third harmonic and unbalancing can be solved by:

(a) Solid grounding of neutrals

(b) Providing tertiary windings

(c) Providing 3-phase 4 wire arrangement

(d) Both (a) and (b)

Ans: **(d)**

Q. 19. In open-delta system, the load that can be carried by the bank without exceeding the ratings of the transformers is X% of the original load carried by the Δ–Δ bank. What is the value of X?

(a) 67.7%   (b) 57.7%

(c) 86.73%   (d) 86.6%

Ans: **(b)**

Q. 20. The main use of Star/Delta connection is:

    (a) At the substation end of the transmission line where voltage is to be stepped down

    (b) At the beginning of high tension transmission system, where it is necessary to step up the voltage

    (c) At the sites, where high voltage, low current transformers are required to be installed

    (d) None of these

Ans: (a)

Q. 21. Open delta (V–V) connection is employed, when:

    (a) The three-phase load is very small

    (b) It is anticipated that in near future, load will increase

    (c) One of the transformers is disabled in Delta–Delta bank and service is continued at reduced capacity

    (d) All of the above

Ans: (d)

Q. 22. What will be the value of power supplied by V–V bank?

    (a) $P_1 = P_2 = $ kVA cos $(30° - \varphi)$

    (b) $P_1 = $ kVA cos $(30° - \varphi)$ and $P_2 = $ kVA cos $(30° + \varphi)$

    (c) $P_1 = $ kVA cos $(90° - \varphi)$ and $P_2 = $ kVA cos $(90° + \varphi)$

    (d) None of the above

Ans: (b)

Q. 23. Two 3 limb, 3-phase Delta–Star connected transformers are supplied from the same source. One of the transformers is of Dy1 and the other is of Dy11 connection. The phase difference between the corresponding phase voltages of the two secondaries would be:

    (a) 0°   (b) 30°

    (c) 60°   (d) 120°

Ans: (c)

Q. 24. A 3-phase balanced sinusoidal rated voltage is fed to the star side of star/delta bank of three single-phase transformers having the same turns ratio. A voltmeter

across an open corner of the secondary delta measures nearly 50% of normal phase voltage. This voltage is due to:

(a) Unequal per unit impedances of the transformers

(b) Equal per unit impedances of transformers

(c) Wrong connections w.r.t. polarity

(d) None of the above

Ans: **(a)**

Q. 25. In a Delta–Star connected transformer, phase to phase voltage transformation ratio is a (delta phase)/1 (star phase). What will be the value of line-to-line voltage ratio in Y–Δ connection?

(a) $a\sqrt{3}/5$

(b) $\sqrt{3}/a$

(c) $a/\sqrt{3}$

(d) $3a^2$

Ans: **(b)**

Q. 26. On which factor, leakage flux in the transformer winding depends?

(a) The mutual flux

(b) Transformation ratio

(c) Applied voltage and frequency

(d) The load current

Ans: **(d)**

Q. 27. The ordinary transformation of three-phase current alone voltage to three-phase current at another voltage can be effected:

(a) By using a single three-phase transformer

(b) By using three separate single-phase transformers

(c) Either (a) or (b)

(d) None of the above

Ans: **(c)**

Q. 28. When a balanced 3-phase load of 40 kVA is supplied, then what will be the rating of each transformer in V–V bank?

(a) 34.6 kVA

(b) 27 kVA

(c) 23 kVA

(d) 25 kVA

Ans: (c)

Q. 29. Which one is the application of open-delta system of transformers?

(a) As a temporary measure, when one transformer of a Δ–Δ system is demaged and removed for maintenance purpose

(b) To provide service in a new development area, where full growth of load may need several years

(c) To supply a combination of large single-phase and smaller 3-phase loads

(d) All of the above

Ans: (d)

Q. 30. What will be utilisation factor for the transformers connected in open-delta?

(a) 0.879   (b) 0.869

(c) 0.667   (d) 0.866

Ans: (d)

Q. 31. In a particular condition of 3-phase transformers, a Δ–Y transformer with delta connection on the primary side operates in parallel mode with a Star–Delta transformer with star connection on the primary side. What will be the value of X, if the ratio N_1/N_2 (per phase) of the former is X times that of the latter?

(a) 3   (b) $\sqrt{3}$

(c) 1/3   (d) $1/\sqrt{3}$

Ans: (a)

Q. 32. The function of multi-stepped core in a transformer is to?

(a) Reduce the cost of copper

(b) Increase the efficiency

(c) Decrease the efficiency

(d) None of the above

Ans: **(a)**

Q. 33. An 11 kV/400 V, 1000 kVA, Y–Y transformer is connected as Delta–Star with the high voltage side connected in delta. The rating for new connection will be?

(a) $11/\sqrt{3}$ kV/400 V, 1000 kVA

(b) $11/\sqrt{3}$ kV/400 V, $1000/\sqrt{3}$ kVA

(c) $11/\sqrt{2}$ kV/440 V, $1000/\sqrt{5}$ kVA

(d) 33 kV/440 V, 1000 kVA

Ans: **(a)**

Q. 34. The transformer tappings are provided at:

(a) The middle of high voltage side

(b) The neutral end of high voltage side

(c) The phase end of high voltage side

(d) The phase end of low voltage side

Ans: **(a)**

Q. 35. A Δ–Δ bank has three 20 kVA, 2300/230 V transformers supplies a load of 40 kVA. When one transformer is removed, then what will be the % of rated load carried by each transformer in resulting connection?

(a) 115.5%

(b) 110%

(c) 120%

(d) 135%

Ans: **(a)**

Q. 36. When a V–V bank of two transformers supplies a balanced 3-phase load of power factor cosϕ, then the two transformer will not have:

(a) Same voltage regulation

(b) Same power

(c) Both (a) and (b)

(d) None of the above

Ans: (c)

Q. 37. A 120 kVA, 6000/400 V, Y–Y, 3-phase, 50 Hz transformer has an iron loss of 1600 watts. The maximum efficiency occurs at 3/4 full load. What will be the transformer efficiency at full load with a power factor of 0.8?

(a) 95.57%   (b) 98.57%

(c) 99%   (d) 97.35%

Ans: (a)

Q. 38. Zero sequence currents can flow from a line into a transformer bank, if the windings are:

(a) Grounded star/delta   (b) Star/grounded star

(c) Grounded star/star   (d) Delta/delta

Ans: (b)

Q. 39. A bank of three identical single-phase 250 kVA, 11 kV/230 V transformers is used to provide 400 V low tension supply from a 11 kV 3-phase substation. The effective kVA rating of bank will be:

(a) 750   (b) 735

(c) 500   (d) $300\sqrt{7}$

Ans: (a)

Q. 40. In case of 3-phase, Star–Star (Y–Y) transformers, distortion in voltage wave shape due to 3rd harmonic flux and voltage unbalance on 1-phase loads are minimum in:

(a) Three-phase three limb core type

(b) Three-phase five limb core type

(c) Three-phase shell type

(d) None of the above

Ans: **(a)**

Q. 41. In case of short circuit test on a transformer, the number and duration of tests on a phase are:

(a) 3 and 0.5 sec ± 10% tolerance

(b) 5 and 2 sec ± 10% tolerance

(c) 1 and 7 sec ±10% tolerance

(d) 1 and 0.5 sec ±10% tolerance

Ans: **(a)**

Q. 42. When a transformer is fed from fundamental frequency voltage source, then the source of harmonics is the:

(a) Saturation of core

(b) Iron loss

(c) Poor insulation facility

(d) None of the above

Ans: **(a)**

Q. 43. When two, 3-phase transformers are operated in parallel, then which of the section will be impossible?

(a) A—Y/Y and B—Δ–Δ

(b) A—Δ–Δ and B—Δ–Δ

(c) A—Y/Δ and B—Δ–Δ

(d) A—Δ/Y and B—Δ–Δ

Ans: **(d)**

Q. 44. Three-phase shell or five limb core type units suffer from the following disadvantage(s):

(a) This cannot supply unbalanced loads between line and neutral

(b) Their phase voltages may get distorted by 3rd harmonic emfs

(c) Both (a) and (b)

(d) None of the above

Ans: **(c)**

Q. 45. A transformer may be subjected to over voltages produced in the line by:

    (a) Faults

    (b) Switching

    (c) Atmospheric conditions

    (d) All of the above

Ans: (d)

Q. 46. Scott connection is used for:

    (a) Three-phase to two phase transformation

    (b) Six phase to twelve phase transformation

    (c) Three-phase to six phase transformation

    (d) None of the above

Ans: (a)

Q. 47. Teaser transformer operates on 0.866 of its rated:

    (a) Current

    (b) kVA rating

    (c) Voltage

    (d) Power

Ans: (c)

Q. 48. In scott connection, the transformers used are named as:

    (a) Teaser transformer only

    (b) Tertiary winding transformer and main transformer

    (c) Both main and teaser transformers

    (d) None of the above

Ans: (c)

Q. 49. In star connection with earthed neutral, the maximum voltage of phase winding to ground is X% of the line voltage. What will be the value of X?

    (a) 59%   (b) 86.6%

    (c) 99.32%   (d) 57.7%

Ans: (d)

Q. 50. Which of the following connections cause interference to the nearby communication systems?

(a) Delta–Star

(b) Star–Star

(c) Delta–Delta

(d) Star–Delta

Ans: **(b)**

MISCELLANEOUS

Q. 1. In a 3-phase transformer, efficiency is more as compared to a bank of three single-phase transformers, because:

(a) It has shorter magnetic path

(b) Core loss is smaller

(c) Both (a) and (b)

(d) None of these

Ans: **(c)**

Q. 2. If the transformer winding has an impedance of 5% under short circuit condition, then it will experience the force in radial direction. What will be its value?

(a) 35 times full load value

(b) Same as in previous case

(c) It will increase to 20 times full load value

(d) It will increase to 400 times full load value

Ans: **(d)**

Q. 3. In an ordinary two winding transformer, 220 V is stepped down to 110 volts. However if an auto transformer is used for the same purpose, then what will be the ratio of copper weights in them?

(a) 2   (b) 0.8

(c) 0.5   (d) 0.7

Ans: **(c)**

Q. 4. Two single-phase transformers, having rating of 100 kVA each were connected in open delta mode. In

this condition, what would be the output capacity of complete arrangement?

(a) 73.2 kVA   (b) 73.9 kVA

(c) 173.2 kVA   (d) 100 kVA

Ans: **(c)**

Q. 5. **In a power transformer, harmonic current does not increase:**

(a) Copper loss

(b) Total loss in transformer

(c) Impedance drop

(d) Secondary voltage

Ans: **(d)**

Q. 6. **By using a variac transformer, output voltage may be varied over a range of .... of input voltage:**

(a) 0 to 10%   (b) 0 to 115%

(c) 0 to 120%   (d) 0 to 50%

Ans: **(c)**

Q. 7. **Which of the following materials is used for insulation purpose?**

(a) Pressboard

(b) Cloth

(c) Mica and impregnating compounds

(d) All of the above

Ans: **(d)**

Q. 8. **In 3-phase transformer, minor insulation includes:**

(a) Insulation between the elements of a given winding such as conductor insulation

(b) Insulation between turns, layers and coils

(c) Both (a) and (b)

(d) Insulation between any two windings

Ans: **(c)**

Q. 9. In 3-phase transformers, which of the following specifications is not met by oil:

(a) Purity of oil

(b) Good resistance to emulsion

(c) High flash point

(d) High viscocity

Ans: **(d)**

Q. 10. In power transformers, the value of flux density and % impedance varies in the range of:

(a) 1.5 to 1.7 tesla and 6% to 18% respectively

(b) 1.7 to 1.8 tesla and 6% to 10% respectively

(c) 1.25 to 1.75 tesla and 3% to 12% respectively

(d) 1.25 to 2.25 tesla and 6% to 30% respectively

Ans: **(a)**

Q. 11. Normally, what is the phase relationship between the primary and secondary voltages of transformer?

(a) 90° out of phase

(b) 180° out of phase

(c) 30° out of phase

(d) 60° out of phase

Ans: **(b)**

Q. 12. In transformers, magnetic leakage problem can be reduced:

(a) By using shell type construction

(b) By arranging both primary and secondary windings concentrically

(c) By sandwitching primary and secondary windings

(d) All of the above

Ans: **(d)**

Q. 13. In which type of transformer, sandwitch type of coils are used?

(a) Shell type

(b) Spiral core or wound core type

(c) Core type

(d) None of the above

Ans: **(a)**

Q. 14. On which basis, autotransformer and two winding transformer may be compared?

(a) Core size and total cost

(b) Leakage impedance and voltage regulation

(c) Requirements of conductor material

(d) All of the above

Ans: **(d)**

Q. 15. Which one is not the factor for production of noise in transformers?

(a) Mechanical vibration of tank walls

(b) Magnetostriction

(c) Mechanical vibrations due to laminations, depending upon the tightness of clamping, size, etc.

(d) All of the above

Ans: **(d)**

Q. 16. Which one of the followings is the disadvantage of Delta–Delta connection?

(a) There is no star (neutral) point available

(b) If a third harmonic is present, it circulates in the closed path

(c) Both (a) and (b)

(d) None of the above

Ans: **(a)**

Q. 17. Three-phase to two-phase conversion is required to supply:

(a) Two-phase furnaces

(b) To link two phase circuit with 3-phase system

(c) 3-phase apparatus from a 2-phase supply source

(d) All of the above

Ans: **(d)**

Q. 18. In scott connection, under unbalanced load, the main transformer rating is X% greater, than that of teaser. What is the value of X?

(a) 15%   (b) 10%

(c) 20%   (d) 7.92%

Ans: **(a)**

Q. 19. An important advantage of scott connection over the V–V connection for 3-phase power transformation is that it provides:

(a) A real 3-phase, 4 wire system

(b) A higher ratio of utilization

(c) More voltages

(d) A set of balanced voltages under load

Ans: **(a)**

Q. 20. Leakage fields present in a transformer induce eddy currents in:

(a) Conductors   (b) Tanks

(c) Channel and bolts   (d) All of the above

Ans: **(d)**

Q. 21. Which of the statements is wrong?

(a) Dielectric loss occurs in insulating material

(b) Stray load loss and dielectric loss is very small

(c) Leakage fields induce harmonic currents

(d) All of the above

Ans: **(c)**

Q. 22. Which one of the statements is wrong w.r.t. the value of insulation resistance (as considered safe with 40°C)?

(a) Rated voltage 400 V, insulation resistance 2 $\mu\Omega$

(b) Rated voltage 33 kV, insulation resistance 150 $\mu\Omega$

(c) Rated voltage 132 kV, insulation resistance 500 $\mu\Omega$

(d) Rated voltage 33 kV, insulation resistance 20 $\mu\Omega$

Ans: **(d)**

Q. 23. What is the maximum permissible operating temperature of class 'E' insulating material?

(a) 150°C

(b) 135°C

(c) 130°C

(d) 120°C

Ans: (d)

Q. 24. Which items are tested in the yearly period of checking of a power transformer?

(a) Sludge formation only

(b) Alarm circuits, contact relays

(c) Earth resistance and acidity in transformer oil

(d) All of the above

Ans: (d)

Q. 25. What must be the dielectric strength of a transformer oil?

(a) 30 kV for a 4 mm gap

(b) 35 kV for a 6 mm gap

(c) 30 kV for a 4.5 mm gap

(d) None of the above

Ans: (a)

Q. 26. Routine tests on a transformer means:

(a) Tests made by the manufacturer on each transformer

(b) Tests made to verify a design characteristics

(c) None of the above

(d) Both (b) and (a)

Ans: (a)

Q. 27. Which one of the followings is not the purpose of tertiary winding?

(a) Tertiary windings are used for supplying substation auxiliaries

(b) These are used for measurement of voltage of high voltage testing transformers

(c) These are used to interconnect three supply systems operating at different voltages

(d) These are used for 3-phase to 2-phase transformation

Ans: **(d)**

Q. 28. **The main application of three-phase autotransformers is for interconnecting two power systems of different voltages such as:**

(a) 66 kV to 132 kV   (b) 132 kV to 1230 kV

(c) 220 kV to 400 kV   (d) Only (a) and (c)

Ans: **(d)**

Q. 29. **Which of the conditions is not true for parallel operation of 3-phase transformers?**

(a) The phase sequence must be the same

(b) The phase shift between primary and secondary voltages must be the same for all transformers, which are being implemented in parallel mode

(c) Both (a) and (b)

(d) The phase sequence must not be the same

Ans: **(d)**

Q. 30. **A 400 kVA load at 0.7 PF lagging is supplied by using three 1 phase transformers, connected in Δ–Δ. Each of the Δ–Δ transformers is rated at 200 kVA, 2300 V/230 V. When one faulty transformer is removed from service, then what will be the value of percent rated load carried by each transformer?**

(a) 115.5%   (b) 110%

(c) 120%   (d) 86.6%

Ans: **(a)**

Q. 31. **In the above example, what will be the percent increase in load on each transformer, when one transformer is removed?**

(a) 73%   (b) 86.6%

(c) 73.2%   (d) 173.2%

Ans: **(c)**

Q. 32. While performing scott connection, the teaser transformer has a primary voltage rating that is at:

(a) $\sqrt{3}/5$ of the voltage rating of main transformer

(b) $1/\sqrt{2}$ of the voltage rating of main transformer

(c) $3/\sqrt{2}$ of the voltage rating of main transformer

(d) $\sqrt{3}/2$ of the voltage rating of main transformer

Ans: (d)

Q. 33. Which one of the connections is not used for 3-phase to six-phase transformation?

(a) Double star

(b) Six-phase star

(c) Star–Delta or Delta–Star

(d) Diametrical

Ans: (c)

Q. 34. In designing transformers, which rating is not preferred as according to BIS 2026?

(a) 160 kVA   (b) 1000 kVA

(c) 25 kVA   (d) 6.5 kVA

Ans: (d)

Q. 35. When two, 3-phase transformers are connected in parallel, then both transformers must have:

(1) Same voltage ratio   (2) Identical polarity

(3) Same phase shift   (4) Same voltage rating

(a) All the data is sufficient

(b) All the data is not essential

(c) Any two of the above are enough

(d) One more essential condition is missing

Ans: (d)

Q. 36. Which one of the systems will have smallest size?

(a) 100 kVA, 11/0.4 kV   (b) 100 kVA, 5.5/0.2 kV

(c) 200 kVA, 22/0.8 kV   (d) 200 kVA, 44/1.6 kV

Ans: (a)

Q. 37. **In a dry type transformer, dust should never be allowed to accumulate on the windings and core because it will:**

(a) Short circuit the windings

(b) Open circuit (break up in the circuit) the windings

(c) Reduce the heat dissipation

(d) None of the above

Ans: **(c)**

Q. 38. **When a high voltage test is applied to the primary winding of a rebuilt transformer, then:**

(a) The secondary, transformer case and core must be grounded

(b) Secondary winding must be open circuited

(c) All the parts of the transformer must be properly grounded

(d) None of the above

Ans: **(a)**

Q. 39. **An autotransformer is mainly used over an two winding transformer due to:**

(a) Operation's safety

(b) Saving in copper, when transformation ratio is near unity

(c) Availability of taps on the secondary

(d) None of the above

Ans: **(b)**

Q. 40. **In an autotransformer, with $K = V_2/V_1$ (as transformation ratio), the fraction of power transferred conductively from the input to the output is:**

(a) K (b) $1 - K$

(c) $1 - 0.5\,K$ (d) $1 + K$

Ans: **(a)**

FILL IN THE BLANKS

1. Three-phase transformers may be of .... type.

Ans: Core type or shell

2. A 3-phase transformer, requires........ part of the iron for the magnetic circuit compared to a bank of three single-phase transformers.

Ans: 1/3rd

3. A 3-phase transformer is ....% cheaper compared to a bank of three single-phase transformers.

Ans: Nearly 15

4. 3-phase transformers are more efficient compared to a bank of three 1-phase transformers, due to the fact that it has .... magnetic path and therefore core volume and core loss is smaller.

Ans: Shorter

5. In transformers, LV winding is placed next to core and HV winding is placed over .... winding with suitable level of insulation between .... and between LV and HV winding.

Ans: Low voltage; The core and LV winding

6. In 3-phase transformer, fluxes created by three windings are .... apart in time phase.

Ans: 120°

7. The .... type construction is generally used in small rating transformers.

Ans: Shell

8. .... coils are used in high voltage windings of large capacity transformers.

Ans: Disc

9. .... windings are preferred in transformers to overcome voltage surges.

Ans: Shielded

10. Cooling arrangement is very essential for .... rating power transformers.

Ans: Large

11. The humming noise in a transformer depends on ....

Ans: Flux density

12. In case of transformers, the use of steel in magnetic circuit introduces iron or core loss, but gives a .... permeability of magnetic circuit.

Ans: High

13. Transformer grade steel consists of .... silicon.

Ans: 3.5%

14. The content of silicon in core affects ....

Ans: Tensile strength and ductility

15. Cold rolled steel is nearly .... expensive compared to hot rolled steel.

Ans: 25%–35% more

16. In transformers, thickness of laminations should not be made below .... because in that condition, the lamination becomes mechanically weak.

Ans: 0.3 mm

17. .... windings are used in core type transformers and .... windings are employed in shell type transformers.

Ans: Concentric; Sandwitch

18. .... windings are made in layers and use rectangular or round conductors.

Ans: Cylindrical

19. Insulation used in the transformer may be into two major groups, namely .... and .... insulation.

Ans: Major; Minor

20. The oil gauge is provided to indicate ....

Ans: Oil level

21. Breather consists of a small container connected to the vent pipe and contains a dehydrating material like ....

Ans: Silica gel crystals impregnated with cobalt chloride

22. Buchhloz relay is basically .... relay.

Ans: Gas operated

23. The purpose of the markings on transformer leads is for ....

Ans: Standardization

24. The area of yoke of a transformer is usually .... more compared to core.

Ans: 15% to 20%

25. Cooling of transformers is more difficult compared to other machines because it has ....

Ans: No rotating part

26. In Star–Star connection of transformers, by connecting primary neutral to generator neutral, the path for return of .... harmonic currents is provided.

Ans: 3rd and 5th

27. Delta-Delta arrangement is used in those systems, which has ....

Ans: Large currents at low voltage

28. Three-phase transformers are divided into........ main groups as according to the phase difference between the corresponding .... voltages on the HV and LV sides.

Ans: Four; Line

29. The primaries and secondaries of any 3-phase transformer can be independently connected in .... mode.

Ans: A Star (Y) or Delta (Δ)

30. The factor, which affects the choice of connections in a 3-phase transformer includes the availability of path for the flow of 3rd harmonic (exciting) and .... currents.

Ans: Zero sequence (fault)

31. Delta–Delta connection is called 0° connection due to the fact that primary and secondary line voltages are .... each other.

Ans: In phase with

32. Delta–Delta connection is basically suitable for both balanced and .... loading.

Ans: Unbalanced

33. Δ–Δ connection has the disadvantage that there is no .... available.

Ans: Star point (neutral point)

34. Y–Y connection is not satisfactory for unbalanced loading in the absence of ....

Ans: Neutral point

35. In any transformer, the magnetising current contains a very large third harmonic component, which is necessary to overcome .... to produce sinusoidal flux.

Ans: Saturation

36. The unbalancing and third harmonic problems of Y–Y connection may be solved by using either solid grounding of neutrals or providing ....

Ans: Tertiary windings

37. In Delta–Star connection, the primary line voltage is .... the primary phase voltage.

Ans: Equal to

38. In Star–Delta connection, the primary line voltage is equal to .... the primary phase voltage.

Ans: $\sqrt{3}$ times

39. In Star–Delta connection, there is a phase shift of 30° lead between respective .... voltages.

Ans: Line to line or phase

40. Y–Δ connection can be obtained by interchanging the primary and secondary roles in .... connection.

Ans: Δ–Y

41. In delta connection, a path is provided for the circulation of .... without the neutral wire.

Ans: Third harmonics and their multiple

42. One of the most popular method of connecting trans-formers is the .... connection.

Ans: Delta/Star

43. Star–Star connected transformers may operate in a satisfactory manner only when the load is ....

Ans: Balanced

44. In a Δ–Δ connected system, if one of the transformer becomes faulty, the remaining transformers will supply the load but at reduced capacity, i.e. about .... % of that of original Δ–Δ bank.

Ans: Nearly 58

45. In case of open delta arrangement, the removal of one transformer would allow the remaining two transformers to carry........ of the load kVA.

Ans: Nearly 2/3rd (66.7%)

46. The cost of one three-phase transformer is less than the cost of three single-phase transformers required to supply the same ....

Ans: kVA output

47. The switchgear arrangement, bus bar structure and other wiring for a three-phase transformer installation are simple compared to three .... transformers.

Ans: Single-phase

48. The low voltage windings are placed nearer the core limb because it is quite easier to .... the low voltage winding from the core compared to high voltage winding.

Ans: Insulate

49. Depending on the phase displacement of the voltages of high voltage and low voltage sides, transformers are classified into groups called ....

Ans: Vector groups

50. For satisfactory parallel operation of transformers, they should belong to the same .... group.

Ans: Vector

51. A Star–Star connected 3-phase transformer can not be connected in parallel with other Star–Delta (mode)

transformer because it may create .... of secondary side of transformer.

Ans: Short circuiting

52. When two, 3-phase transformers are connected in parallel, then it is necessary that they must have .... phase sequence.

Ans: Same

53. In 3-phase transformers, phase sequence must be .... for parallel connected transformers, otherwise during the cycle, each pair of phases will be ....

Ans: Identical; Short circuited

54. In a scott connected transformer, the number of primary and teaser turns respectively are ....

Ans: N, $\sqrt{3}N/2$

55. In scott connection, the neutral point divides the teaser winding in the ratio ....

Ans: 1:2

56. Scott connection connects two single-phase transformers to perform the 3-phase to .... conversion.

Ans: Two phase

57. In scott connection, one transformer is called as .... and other is called as ....

Ans: Main transformer; Teaser transformer

58. Scott connection may be used to supply 1-ϕ loads, such as electric trains, which are so scheduled as to keep the load on the 3-phase system as nearly .... as possible.

Ans: Balanced

59. The capacity to rating ratio in T–T connection is 86.6%, i.e. the same as in ...., if two identical units are used.

Ans: V–V connection

60. In scott connection, ratio of kVA utilized to that available would be ...., which makes this connection more economical compared to open delta with its ratio of ....

Ans: 0.928; 0.866

61. In 3-phase transformers, Star–Star connection is rarely used in practice due to ....

Ans: **Oscillatory neutral problems**

62. In a 3-phase transformer, third winding is called as .... winding.

Ans: **Tertiary**

63. In 3 winding transformers, kVA ratings of the three windings are normally ....

Ans: **Unequal**

64. Tertiary windings are connected in ....

Ans: **Delta**

65. The important advantages of 3 winding transformer are economy of construction and greater ....

Ans: **Efficiency**

66. Polarities of a transformer identify the relative directions of .... in the two windings.

Ans: **Induced voltages**

67. Polarities may be checked by a simple test requiring only voltage measurements with transformer at .... condition.

Ans: **No load**

68. In 3-phase autotransformers, delta connections are usually avoided and .... are normally used.

Ans: **Star connections**

69. In an autotransformer, there is a loss of .... between input and output circuits.

Ans: **Isolation**

70. The ratio of copper-weight on an autotransformer to that on an ordinary two winding transformer is given as ...., where K is the transformation ratio.

Ans: **1-K**

71. An autotransformer consists .... winding with taps taken out.

Ans: **One**

72. An autotransformer may be utilized as an .... transformer.

Ans: **Voltage regulating**

73. The purpose of an autotransformer may be to start .... motors.

Ans: **Induction and synchronous**

74. In a transformer, overcurrents affect both ....

Ans: **Insulation life and mechanical stresses**

75. The most common cause of acidity in transformer oil is .... of transformer oil.

Ans: **Overheating**

76. The maximum permissible operating temperature of class E insulating material is ....

Ans: **120°C**

77. It becomes necessary to cool the transformers to dissipate .... in the windings.

Ans: **Heat generated**

78. The additional condition for parallel operation of three-phase transformers over single-phase transformers is that the transformers must belong to .... group.

Ans: **Same vector**

79. In marking the terminals, the higher voltage winding is designated by a capital letter and the lower voltage winding by the ....

Ans: **Small letter**

80. A fresh transformer oil must have IFT value of ....

Ans: **0.04 N/m**

81. In transformer, the oil is put air-tight in the transformer tank and installed silica gel breather, inholes .... only at the time of need.

Ans: **Dry air**

82. The iron core tends to ...., when it is subjected to magnetisation reversed cycles due to .... supply, thus producing humming sound.

Ans: **Vibrate; AC**

83. In case of power transformers, additional cooling arrangements are provided so as to increase .... rating and to provide longer life to ....

Ans: **Power; Insulation**

84. The power transformers used may be of indoor, outdoor and .... type.

Ans: **Pole mounted**

85. The voltage transformation ratio for an autotransformer should be nearly ....

Ans: **Unity**

86. The advantage of using stepped core in power transformers is the .... in winding material.

Ans: **Economy**

87. In transformer design, higher flux density is used to reduce the .... per kVA.

Ans: **Weight**

88. Star-Star connected transformer is suitable only for .... loads.

Ans: **Balanced**

89. Delta–Delta connected transformer is economical for large current and low voltage systems, in which .... problem is not so urgent.

Ans: **Insulation**

90. In Star–Delta transformer, a sinusoidal flux is obtained due to the flow of .... current in the delta.

Ans: **Third harmonic**

91. The product of the rated voltage in kV, the rated current in amperes and the appropriate phase factor is called as .... of transformer.

Ans: **Rated kVA**

92. A statement of operating limitations assigned to the transformer by the manufacturer under specified conditions is called as .... of transformer.

Ans: **Rating**

93. An auxiliary delta-connected winding, used particularly in star-connected transformer is also called as stabilising or ....

Ans: **Tertiary winding**

94. For large transformers, the tanks of transformers are evacuated to approximately .... mm of mercury.

Ans: **700**

95. .... insulation is used as the inter-turn insulation in low voltage transformers.

Ans: **Enammel**

96. The third winding in a transformer, if provided is known as .... winding.

Ans: **Tertiary**

97. A transformer must be filled with oil through the bottom drain valve to prevent .... of the oil.

Ans: **Aeration**

98. The line current corresponding to the rated kVA and the rated voltage is called as .... of a transformer.

Ans: **Rated current**

99. Two main advantages of running transformer in parallel are optimum .... of load and reduction in ....

Ans: **Sharing; Losses**

100. One disadvantage of 3-phase transformer is that in case of fault condition, the whole system has to be .... for repairs.

Ans: **Shutt off**

TRUE / FALSE

1. Three-phase transformers may be used either to step up or step down the three-phase voltage.

Ans: **True**

2. Switchgear and structure of bus bars for a 3-phase installation are comparatively simpler than those for three single-phase transformers.

Ans: **True**

3. Three 1-φ transformers can be used as a 3-phase transformer.

Ans: True

4. A 3-phase transformer is less efficient compared to a bank of 3 single-phase transformers.

Ans: False

5. Whenever balanced 3-phase sinusoidal voltages are applied to the windings, then the fluxes ϕ_a, ϕ_b and ϕ_c may or may not be sinusoidal.

Ans: False

6. In 3-phase transformers, high voltage winding is placed near the core limb.

Ans: False

7. There exists a specific time phase relationship between the terminal voltages of the high voltage and low voltage sides.

Ans: True

8. If in case of a 3-phase transformer, primary and secondary windings are connected in star and delta modes respectively, then the phase displacement will be 180°.

Ans: False

9. There exists mainly 3 groups as according to the phase difference between the corresponding line voltages on the HV and LV sides.

Ans: False

10. Transformers may be classified into groups called as 'vector groups', according to the phase difference in voltages of HV and LV sides.

Ans: True

11. Parallel operation of 3-phase transformers depend upon the phase displacement between the line voltages of low voltage and high voltage sides.

Ans: True

12. A Star-Star connected, 3-phase transformer may not be connected in parallel mode with other Star–Delta transformer.

Ans: **True**

13. In Δ–Δ transformer, no problem is associated regarding unbalanced loads or harmonics.

Ans: **True**

14. Δ–Δ connection may be proved useful in case of large current and low voltage transformers.

Ans: **True**

15. In Star–Star connection, there is a phase shift of 0°.

Ans: **True**

16. When load is unbalanced and neutral is not provided in 3-phase transformer connection, then the phase voltages tend to become severely unbalanced.

Ans: **True**

17. When there is no path for 3rd harmonic component of current in ungrounded Star–Star connection, then obviously it will try to distort the flux wave.

Ans: **True**

18. Star–Delta connection of transformer is made for a phase shift of 30° lag.

Ans: **True**

19. Y–Δ connection faces problems with unbalanced loads and third harmonics.

Ans: **False**

20. A 3-phase transformer continue to function, in case of fault on one of the transformers.

Ans: **True**

21. If leakage impedances are considered negligible, then the balanced 3-phase line voltages applied to the V–V primaries produce balanced 3-phase voltages on secondary side.

Ans: **True**

22. In case of Star–Star transformers, distortion on voltage wave shape due to 3rd harmonic flux and voltage unbalance on single-phase loads are minimum if they work as banks of 1-phase transformers.

Ans: **False**

23. In case of V–V bank, total VA rating would be equal to $\sqrt{3}$ times, VA rating per transformer in V–V bank.

Ans: **True**

24. In open delta system, each of the transformers will be overloaded by 73.2%.

Ans: **True**

25. Open-delta connection may supply a load combination of single-phase and smaller 3-phase system load.

Ans: **True**

26. In scott connection, 3 single-phase transformers are needed with 86.6% tapping on the primary side.

Ans: **False**

27. In case of balanced load with power factor $\cos\theta$ (lag.), teaser transformer operates at $\cos\theta$ lag., but two halves of main transformer operate at a power factor of $\cos(30° + \theta)$ and $\cos(30° - \theta)$.

Ans: **True**

28. When the teaser transformer is designed for line voltage (V) with taps on both the sides at 86.6% points, 13.4% of the winding will be idle.

Ans: **True**

29. 3-phase to 2 phase conversion may be useful, to supply three-phase load from two phase system or to interlink 2φ and 3φ systems.

Ans: **True**

30. In scott connection, if the two phase side is connected to a balanced two phase supply, the three-phase side will produce balanced three-phase voltage.

Ans: **True**

31. In three-phase to 2 phase conversion, various connections such as Taylor connection, Fortesue connection and Arnold connection may be used.

Ans: True

32. The teaser transformer may also be called as auxiliary transformer.

Ans: True

33. The neutral point on teaser winding divide it in the ratio of 3:1.

Ans: False

34. If the 2-phase currents are balanced, the three-phase currents will not be balanced.

Ans: False

35. For a 2-phase balanced load condition, the phase difference (φ) between teaser secondary voltage and current will be same as the phase difference between main secondary voltage and current.

Ans: True

36. Generally 2-phase generators are not available, so the conversion from 2-phase to 3-phase is not used in practice.

Ans: True

37. 3φ to 6φ conversion may be proved useful in Thyristor and rectifier circuits.

Ans: True

38. In double delta connection of 3φ to 6φ conversion, three identical transformers, each with two independent but identical secondaries are required.

Ans: True

39. In reality, when the six terminals are connected to a suitable 6-phase load, then only a true 6-φ supply may be obtained.

Ans: True

40. In double delta connection, two secondary deltas provide the required path for third harmonic current.

Ans: True

41. In Star–Delta double star connection (for 3-phase to 12-phase), two banks of three transformers or two three-phase transformers are needed.

Ans: True

42. Whenever a third winding is added in transformer, it may be called as tertiary winding.

Ans: True

43. Tertiary winding may be connected in star mode.

Ans: False

44. In triple wound transformers, kVA ratings of the three windings are usually unequal.

Ans: True

45. Star connection may suppress any harmonic voltage, which may be generated in star connected primaries of transformer.

Ans: False

46. Tertiary winding may be employed for measuring voltage of high voltage testing transformers.

Ans: True

47. Triple wound transformers may be manufactured with tertiary VA ratings up to 35% of total VA rating of the transformer.

Ans: True

48. Triple wound transformers give highest possible efficiency.

Ans: True

49. By using star connected tertiary winding, zero sequence currents may be provided low reactance paths.

Ans: False

50. Polarities of transformers can be checked by a simple test requiring only voltage measurements with transformer on no load.

Ans: True

51. The polarity may be either of positive or negative nature.

Ans: True

52. The parallel operation of 3-phase transformers may be useful, when a transformer is taken out of service for its maintenance and inspection.

Ans: True

53. All transformer's terminals must be connected to a particular (in parallel mode) bus, so that they may have same polarities.

Ans: True

54. The phase sequence (order in which the phases attain their maximum voltage) must be different for two parallel banks of transformers.

Ans: False

55. When a third winding is added to a Y–Y connection, then this connection may be known as Y–Y–Y connection.

Ans: False

56. Circulating currents are produced between the transformers operating in parallel, due to unbalanced voltage.

Ans: True

57. Circulating current increases the permissible output of the bank.

Ans: False

58. The three-phase transformer calculations may be made on a per phase basis under balanced loaded conditions.

Ans: True

59. For all those transformers, which are being connected in parallel, one important condition is that the phase shift between primary and secondary voltages must be the same.

Ans: True

60. In case of three-phase autotransformers, star connections are normally avoided.

Ans: False

61. When applied voltage to primary winding is sinusoidal, then no load current I_0 will be also sinusoidal due to hysteresis loop.

Ans: False

62. No load current waveshape contains only 3rd and 5th harmonics.

Ans: False

63. Whenever the transformer is switched on initially, then a very sudden inrush of current takes place in primary winding.

Ans: True

64. In some special cases, the effective value of magnetising current might be larger than the primary rated current of the transformer.

Ans: True

65. When the transformer is connected to the supply line, near a positive or negative voltage maximum, the current inrush will be minimized.

Ans: True

66. To produce a sinusoidal flux, the phase magnetising currents must contain 3rd and higher harmonics.

Ans: True

67. A transformer may be subjected to overvoltages produced in the line by switching, faults, atmospheric conditions such as lightning discharges, etc.

Ans: True

68. When the primary winding of transformer is switched on at a point in the near zero region of the voltage waveform, very high initial current surge may occur.

Ans: True

69. Transformer insulation breakdown under surges may be avoided by adopting fibre shields.

Ans: False

70. Surge absorbers reduce the sleepness of the wave front and the eddy losses in the tank dissipate surge energy appreciably.

Ans: True

VIVA VOCE QUESTIONS

Q. 1. What material is used for making the core of transformer?

Ans: Silicon steel is preferred.

Q. 2. What do you mean by a 3-phase transformer?

Ans: It is that transformer, which is equivalent to three single-phase transformers but wound on one core and enclosed within one common base.

Q. 3. Why the humming noise is produced in a 3-phase transformer?

Ans: The iron core tends to vibrate, when it is subjected to reversed cycles due to AC supply, thus producing humming sound. By tightening the core, it can be reduced.

Q. 4. What do you mean by stabilizing winding of a transformer?

Ans: It is a winding, which may be designed to supply small external load. It may also be used to limit short circuit currents in the circuitry.

Q. 5. What are the advantages of a 3-phase unit transformer?

Ans: It has following advantages:
(a) It takes less space
(b) It is slightly (nearly 15%) more efficient
(c) The costly high voltage terminals to be brought out of the transformer housing are reduced to three rather than six necessary for three separate single-phase transformers
(d) The bus bar structure, switchgear and wiring for a single 3-phase transformer installation are simpler than those for a transformer bank.

Q. 6. What do you mean by grouping in case of a 3-phase transformer?

Ans: Three-phase transformers are divided into four main groups according to the phase difference between the corresponding line voltages between HV and LV sides. It is called as grouping.

These groups are:

Group 1—0° phase displacement

Group 2—180° phase displacement

Group 3—(–30)° phase displacement

Group 4—(+30)° phase displacement

Q. 7. What are the four possible connections for a 3-phase transformer bank?

Ans: It includes:

(a) Δ–Δ (Delta in primary–Delta in secondary)

(b) Y–Y (Star primary–Star secondary)

(c) Δ–Y (Delta primary–Star secondary)

(d) Y–Δ (Star primary–Delta secondary)

Q. 8. Mention all those factors, which affect the choice of connections in 3-phase transformers?

Ans: Some of the factors are as given below:

(a) Voltage stress and economic considerations

(b) Operation under fault/emergency conditions

(c) Availability of a neutral connection for grounding, protection or load connections

(d) Path availability for the third harmonic currents

Q. 9. What do you mean by Δ–Δ connection?

Ans: It simply means that both primary and secondary windings are connected in the system in delta (closed path) made.

Q. 10. Under balanced conditions, what is the relationship between line and phase currents?

Ans: In above case, the line currents are $\sqrt{3}$ times the phase (winding) currents and displaced behind the phase currents.

Q. 11. What are the advantages in star connection over delta connection?

Ans: Each star connected transformer is wound for only 57.7% of line voltage. In HV transmission, it admits much smaller transformers, being built for high voltage than possible with the delta connection, due to less insulation.

Q. 12. What are the advantages associated with Δ–Δ transformation?

Ans: The advantages of Δ–Δ connection are:
   (a) However if third harmonic is present, then it will circulate itself in closed path and does not appear in the output voltage wave.
   (b) In case of failure of any of the transformers, two transformers remaining will supply three-phase power without interruption. It is called as open delta (V–V) connection.

Q. 13. What is the basic disadvantage in Δ–Δ connection?

Ans: It has the important disadvantage that there is no star point (neutral point) available.

Q. 14. By which methods, the unbalance and third harmonic problems of Y–Y connection, may be solved?

Ans: By using
   (a) Solid grounding of neutrals or
   (b) Providing tertiary windings

Q. 15. Why the Δ–Y connection or Y–Δ connection has no problem with unbalanced loads and third harmonics?

Ans: Since in these cases, the delta connection assures balanced phase voltages on the Y-side and provides a path for the circulation of 3rd harmonics and their multiples without the use of neutral wire.

Q. 16. What do you mean by open delta (V–V) system?

Ans: When one of the transformers of a Δ–Δ system is accidently opened, the system will continue to supply 3-phase power, when this defective transformer is disconnected and removed, the remaining two transformers continue to work as a 3-phase bank with rating reduced to about 58% of that of the original Δ–Δ bank. It is called as open-delta or V–V system.

Q. 17. How much amount of load will be carried by the V–V bank?

Ans: The VA supplied by each transformer in a V–V system will be 57.7% of total VA.

Q. 18. Why the load must be reduced by $\sqrt{3}$ times in case of an open-delta connected transformer?

Ans: When three transformers in Δ–Δ are supplying rated load and as soon as it becomes a V–V transformer, the current in each phase winding in increased by $\sqrt{3}$ times. It means that full line current will flow in each of the two phase windings of transformers. Thus each of the transformers in V–V system is overloaded by 73.2%. Due to this reason, load must be reduced as according to the above mentioned case.

Q. 19. When a V–V bank supplies load to system, then what will be the powers of both the transformers?

Ans: In this case, when PF is cosφ, then the angle between line voltage and line current in one transformer is (30° + φ) and the angle between the line voltage and line current in other transformer is (30° − φ). Thus one transformer operates at a power factor of cos(30° + φ) and other at cos(30° − φ). It can be represented as

$P_1 = V_L I_L \cos(30° + φ)$, $P_2 = V_L I_L \cos(30° − φ)$

Q. 20. When a bank of two single-phase transformers are connected in an open-delta arrangement to supply a 3-phase load, do they supply their rated output?

Ans: Each transformer is only capable of supplying 86.6% of its output rating.

Q. 21. What are the various applications of open-delta (V–V) system?

Ans: It may be used in the following circumstances:
(a) To provide service in those areas, where full growth of load may require several years
(b) To supply large single-phase loads as well as smaller 3-phase loads
(c) As a temporary circuitry, when transformer of a Δ–Δ system is demaged and taken out of the service for repair.

Q. 22. Which one is the common method for 3-phase to 2-phase conversion?

Ans: The scott connection is the most common method of connecting two single-phase transformers to have 3-phase to 2-phase conversion.

Q. 23. In which mode, the two transformers are connected in above connection?

Ans: The two transformers are connected electrically but not magnetically.

Q. 24. How many transformers are employed in scott connection?

Ans: Two transformers are used, named as main transformer and teaser transformer. Main transformer is a centre-tapped transformer and the teaser transformer has 86.6% tapping.

Q. 25. In which ratio neutral point N divides the primary of the teaser transformer?

Ans: 2:1

Q. 26. At which places, scott connection may be proved useful?

Ans: This connection may be used in

(a) Electric furnaces, where it is required to operate two 1-φ furnaces together and draw a balanced load from a 3-phase supply.

(b) To connect a 3-phase system with a 2-phase system.

Q. 27. For 3-phase to 6-phase transformation, which connections are used?

Ans: These may include:

(a) Double star   (b) Double delta

(c) Six-phase star   (d) Diametrical

Q. 28. In which mode, tertiary winding is connected in the systems?

Ans: The tertiary winding is generally used in delta mode.

Q. 29. What is the important advantage of using tertiary winding?

Ans: The main advantage of using tertiary winding is that the delta connection suppresses any harmonic voltages, which may be generated in star-connected primaries and secondaries of transformers.

Q. 30. What do we mean by triple wound transformers?

Ans: In transformers, a third winding known as tertiary may be provided in addition to primary and secondary winding. Such transformers are called the triple wound (3 winding) transformers.

Q. 31. What are the uses of tertiary winding?

Ans: This winding may serve a number of purposes, i.e. it may be

(a) To interconnect, three supply systems operating at used different voltages.

(b) To supply condensers and substation auxiliaries at a different voltage compared to primary and secondary windings

(c) In star–star connected transformers, to allow earth fault current or magnetising current, for suppressing harmonic voltages and limiting voltage unbalance.

Q. 32. Why the tertiary winding is known as auxiliary or stabilising winding?

Ans: It is called as auxiliary winding because it may be used for supplying an additional small load at different voltage. Similarly it also limits the short circuit current, so called as stabilising winding also.

Q. 33. Why the tertiary windings are delta connected in normal situations?

Ans: Actually in the conditions of fault and short-circuiting on the HV and LV sides, unbalancing is produced in phase voltages which may be compensated by the circulating currents flowing in the Δ path. That's why these windings are connected in closed Δ path.

Q. 34. The value of reactance of tertiary winding must be high, why?

Ans: In reality, to limit the circulating currents or to avoid overheating, reactance must be high.

Q. 35. Why the low voltage and medium voltage windings are kept near to each other?

Ans: It is done so as to reduce the leakage reactance between this pair of windings.

Q. 36. What are the possible arrangements for winding?

Ans: When we start from core outwards, then we have
(a) LV, MV and HV   (b) MV, LV and HV

Q. 37. What are the responsible factors, which may produce noise in transformers?

Ans: These factors may be:
(1) The mechanical vibration of tank walls
(2) Magnetostriction
(3) Damping, etc.

Q. 38. In case of a transmission system, the star side of a star/ delta transformer is HV side, while in a distribution system, the star side is the LV side, why?

Ans: In the former case, a grounded neutral is needed on the high voltage side so as to provide protection for line to ground fault. Similarly in the 2nd case, a grounded neutral and the neutral line is needed on the LV side to feed 1-ϕ loads. That's why the above arrangement is used.

Q. 39. By using Sumpner's test on a 3-phase transformer, which type of information may be obtained?

Ans: By this test, efficiency, regulation and heating under loaded conditions may be obtained.

Q. 40. What is the basic meaning of parallel operation of transformers?

Ans: It is that arrangement in which the primaries of two or more transformers are supplied from the same source and secondaries are connected to supply the same load.

Q. 41. What are the basic conditions, which are required by the two transformers, when they are connected in parallel?

Ans: For satisfactory parallel operation of 3-phase transformers, the necessary conditions are:

(a) Percent impedances of the transformers must be same in magnitude and must have the same phase angle, (i.e., X/R ratio should be equal for both transformers)

(b) The phase sequence of transformers must be same

(c) The polarities of two transformers should be same

(d) The phase displacement between primary and secondary line voltages of the transformers must be equal, etc.

Q. 42. What is the basic difference between ordinary two winding transformer and an autotransformer?

Ans: In a conventional two winding transformer, the primary and secondary windings are completely insulated from each other but are magnetically linked by a common core.

But in an autotransformer, the two windings, primary and secondary are connected electrically as well as magnetically, in fact a part of the single continuous winding is common to primary and secondary.

Q. 43. What are the two basic types of autotransformer?

Ans: Actually in one type of autotransformers, continuous winding is used, with taps brought out at convenient points, determined by the designed secondary voltages and in other type of autotransformers, two or more coils are connected to form a continuous winding.

Q. 44. On which basis, autotransformer and two winding transformer may be compared?

Ans: The factors are:
  (a) Core size
  (b) Voltage regulation and leakage impedance
  (c) Total losses and efficiency
  (d) Requirements of conductor material

Q. 45. What are the applications of autotransformer?

Ans: The applications are as follows:
  (a) In electrical testing laboratory
  (b) As boosters to raise the voltage, wherever required
  (c) For interconnecting systems that are operating at roughly the same voltage
  (d) For starting induction and synchronous motors

Q. 46. When two transformers are connected in parallel mode, what factors must be taken into consideration?

Ans: Their electrical characteristics, such as voltage ratio, percent impedance, polarity, phase sequence etc. must be considered.